Construction Fleet Inventory Guide

Transportation and Regional Programs Division
Office of Transportation and Air Quality
U.S. Environmental Protection Agency

Prepared for EPA by
Eastern Research Group, Inc., subcontractor to Abt Associates, Inc.
EPA Contract No. EP-W-06-044
Task Order No. 57

NOTICE

This technical report does not necessarily represent final EPA decisions or positions. It is intended to present technical analysis of issues using data that are currently available. The purpose in the release of such reports is to facilitate the exchange of technical information and to inform the public of technical developments.

United States
Environmental Protection
Agency

EPA-420-B-10-025
July 2010

Table of Contents

List of Figures

Introduction

With an increased focus on air quality in this country, there are more and more incentive programs available to diesel equipment and vehicle owners to encourage them to replace older equipment with newer, cleaner ones, to rebuild old engines, or to retrofit current engines with new technologies that help lower emissions. However, in order to apply for funding from many of these programs, you must collect detailed information on your current equipment and/or vehicle(s). Gathering all of the necessary information for program application can prove daunting, as the information requested is not always readily available. Information of interest may include, among others:

- Equipment or Vehicle Type,
- Engine Make,
- Engine Model,
- Engine Model Year,
- Engine Displacement,
- Engine Horsepower,
- Engine Tier Level, and
- Annual Activity.

Purpose of Guide

This guide will aid you in collecting the information you will need to create an accurate inventory for your construction fleet. The information you collect from your vehicles/equipment, with the aid of this guide, will assist you in assessing your fleet for its ability to qualify for various clean diesel incentive programs. This guide also outlines best practices for collecting the required information and removing some of the barriers to finding and identifying the required information about your fleet. Using this guide to collect vital information about your fleet will aid you in identifying ways to make your fleet "cleaner" through volunteer measures, or to meet state and federal mandates.

How to Use This Guide

To compile your inventory data, answer the questions in the "Requirements" section below. If you have trouble answering any of the questions, refer to the appendices for more information and descriptions on the vehicles and equipment in your construction fleet. The "Helpful Resources" section also provides valuable references that may help answer the questions in the "Requirements" section.

Limitations to this Guide

While an extensive effort has been made to provide pictures and descriptions of a wide variety of equipment and vehicles types, and to specify the general location where the required information can be found, the exact location of the information and engine configurations can vary greatly between different makes, models, and model years. As such, this guide is not intended to represent every make, model, or model year specifically, only to provide general guidance in identifying and locating the information of interest. If, after referring to this guide, you are still unable to identify key information about your equipment/vehicle, you may contact the manufacturer for information. Contact information for many construction equipment manufacturers is provided in Appendix A.

Requirements for Building an Inventory

Exactly what information do you need to build an accurate inventory of your construction fleet? The following questions will help you gather all of the information you will need about your fleet. Of the information you will be collecting, some information is more critical than others. The table below highlights the most important information, common to almost all incentive programs, necessary for building your inventory. The remaining information you collect may be required by some incentive programs and not others. This guide could also differentiate the pieces of information needed by engine companies for engine upgrades and repowers and retrofit manufacturers to match engines and applications to the appropriate retrofits.

<u>**Critical Inventory Components**</u>

Equipment/Vehicle Application	
Annual Activity (miles per year for on-road vehicles, hours per year for off-road equipment)	
Engine Manufacturer	
Engine Model	
Engine Model Year	
Fuel Type	
Gross Vehicle Weight Rating for On-road Vehicles	

Engine Horsepower
Engine Displacement

Answer the following questions to build your inventory.

1. **What is the major application for the vehicle/equipment?**
   a. On-road
   b. Off-road

2. **What is the fleet type?**
   a. Public
   b. Private
   c. Rental

3. **What is the primary use of this equipment?[1]**
   a. On-road vehicles
      i. Short-haul
      ii. Long-haul
      iii. Delivery truck
   b. Off-road
      i. General construction

---
[1] Not all of these will apply specifically to the construction industry.

    ii. Road paving

   iii. Land clearing/Excavation

   iv. Digging/Trenching

4.   What type of vehicle or equipment is it?

  a. On-road vehicles

To help identify the type of on-road vehicle, first determine the Gross Vehicle Weight Rating (GVWR). The GVWR for a vehicle is the maximum allowable total weight of a vehicle, or the weight that equals the total unladen weight of the vehicle plus the heaviest load that can be transported by the vehicle. The GVWR can sometimes be found in the owner's manual for the vehicle. Otherwise, most on-road vehicles have a sticker with the GVWR information posted in one of several places:

- Driver's side door

- Driver's side door frame

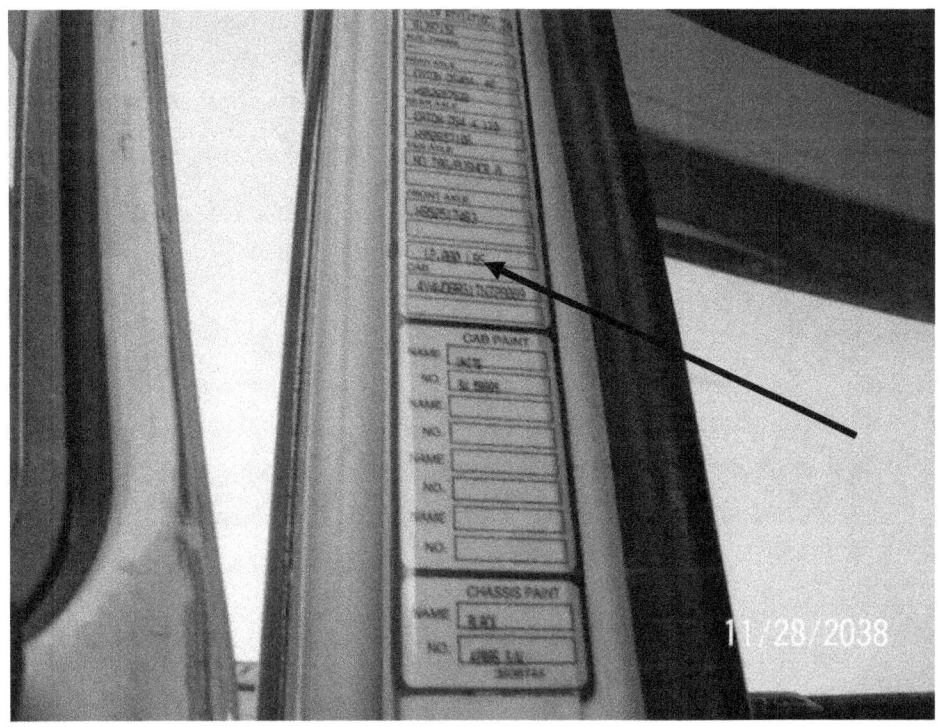

- Under the hood near the radiator

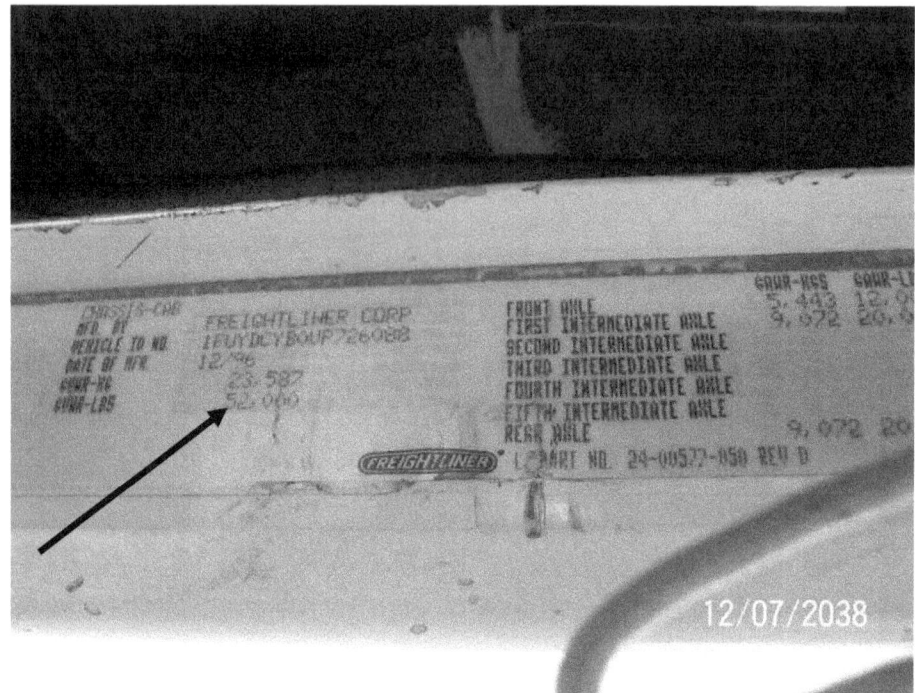

- Printed on the outside of the vehicle (for commercially-licensed vehicles)

The table below shows the vehicle classifications based on GVWR. Examples of types of truck included in each of the vehicle classifications can be seen in Appendix B.

GVWR Heavy-Duty Truck Classifications[2]

	Class	Descripton	GVWR (lbs)
DIESEL	HDDV2B	Class 2b Heavy-Duty Diesel Vehicles	8,501-10,000
	HDDV3	Class 3 Heavy-Duty Diesel Vehicles	10,001-14,000
	HDDV4	Class 4 Heavy-Duty Diesel Vehicles	14,001-16,000
	HDDV5	Class 5 Heavy-Duty Diesel Vehicles	16,001-19,500
	HDDV6	Class 6 Heavy-Duty Diesel Vehicles	19,501-26,000
	HDDV7	Class 7 Heavy-Duty Diesel Vehicles	26,001-33,000
	HDDV8A	Class 8a Heavy-Duty Diesel Vehicles	33,001-60,000
	HDDV8B	Class 8b Heavy-Duty Diesel Vehicles	>60,000

b. Off-road equipment

There is a wide variety of off-road construction equipment. It is critical to appropriately identify your equipment because in most cases, retrofits are both equipment type- and horsepower-specific. If you are unsure what type of equipment you are collecting information on, Appendix C provides descriptions of some of the most common types of construction equipment.

5.   What is the annual activity for this vehicle/equipment?

a. On-road vehicles

How many miles does this vehicle travel in a year? Maintenance records on this vehicle should help you identify how many miles this vehicle travels in a year. If maintenance records are unavailable, record the total number of miles traveled by this vehicle from the odometer, located on the dash instruments. To estimate the average annual mileage, divide the total miles by the age of the vehicle.

b. Off-road equipment

How many hours is this equipment used in a year? Again, maintenance records for this equipment should help you identify how many hours the equipment is operated in a year. If maintenance records are unavailable, record the total number of hours operated from the hour-meter on the equipment. The hour meter may be located on the dash instruments of the equipment as seen in Figure 1, or may be inside the engine compartment as seen in Figure 2. To estimate the average annual operating hours, divide the total number of hours by the age of the equipment.

---

[2] Vehicle weight classifications and descriptions are from EPA's MOBILE6 model.

Furthermore, hour meters are often broken on older equipment or may not be present at all and maintenance records may not be well-kept. In these instances, you may use fuel usage as a means to estimate activity. However, fuel usage will depend greatly on the load the equipment operates under. For example, a bulldozer actively performing land clearing will consume more fuel than the same bulldozer sitting at idle. You should take this into consideration when using fuel logs to estimate activity. Appendix D illustrates the average fuel consumption (in gallons per year) used by a single piece of diesel construction equipment, by horsepower range and assumes a single unitworks 1,000 hours/year.[3]

Figure 1. Hour Meter on Dash Instruments

---

[3] Average fuel consumption examples were generated with the U.S. EPA NONROAD2008a model, using default load factors and emission factors. The model run used 2010 as the episode year and assumed all other inputs were default, except population, which was set to one for all equipment types and horsepower ranges, and activity, which was set to 1,000 hours for all equipment types.

Figure 2. Hour Meter Inside Engine Compartment

6. **Where does the activity occur for this vehicle/equipment?**

For many state-run incentive programs, the location where the activity occurs is very important for identifying potential emissions reductions in targeted areas facing air quality challenges. For many programs, you must be able to report the percentage of annual activity, either miles for on-road vehicles or hours for off-road equipment, in a particular location, usually on a by-county basis.

7. **What are the make, model, and model year of the vehicle/equipment <u>chassis</u>?**
    a. On-road vehicles

    Each on-road vehicle has a unique Vehicle Identification Number (VIN). The VIN conveys a wealth of information about the vehicle, including make, model, and year of manufacture.

Every on-road vehicle will have a VIN affixed in one (or more) of several places:

- Driver's side dash at the base of the window

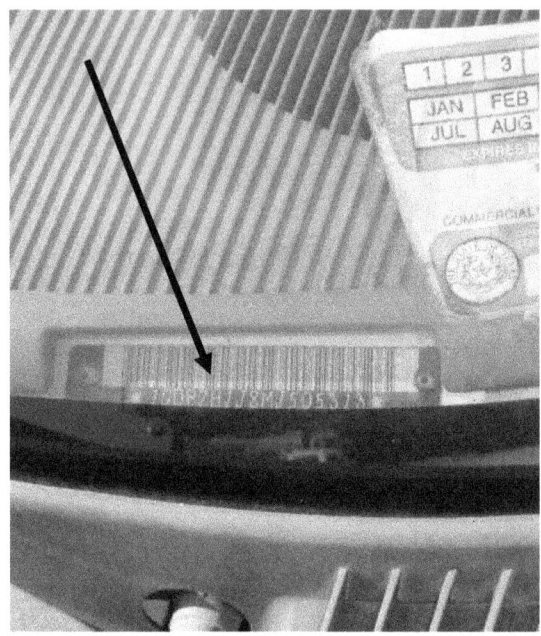

- On the inside of the driver's door

- On the edge of the door

- On the base of the driver's seat

- On door jamb

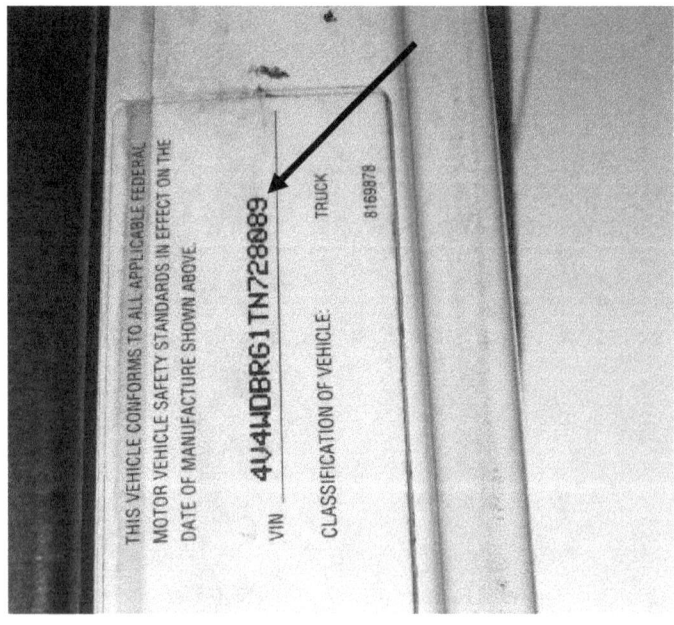

- On the underside of the dashboard on the driver's side

- On the firewall near the clutch pedal

Beginning in 1981, the VIN was standardized to a 17-digit format and will not contain the letters I, O, or Q, to prevent confusion with the numbers 1 or 0. For more information on how to decode a VIN to obtain the required make, model, year and other information of interest, please refer to Appendix E.

a. Off-road equipment

Off-road equipment usually has the make and model clearly labeled and highly visible on the exterior of the equipment. Also, once you identify the serial number, you can contact your dealership, or visit their Internet site, to obtain the required information specific to your equipment. However, without the engine serial number or purchase records, you may not be able to identify the model year.

2. What is the EPA engine family name?[4, 5]

The engine family name is a 12-digit alpha-numeric code used by the U.S. EPA to classify vehicles and engines for the purpose of emissions certification. An engine may have an exhaust engine family name and an evaporative engine family name, depending on the year the engine was manufactured. The engine family name conveys a wealth of information about the engine. The engine family name is located on a label or plate in the engine compartment in a visible position such as the hood underside, shock tower, radiator support, fan shroud, or firewall. Some examples of engine labels inside of the engine compartment are shown in Figures 3 through 6.

Figure 3. Engine Family Label Inside the Engine Compartment

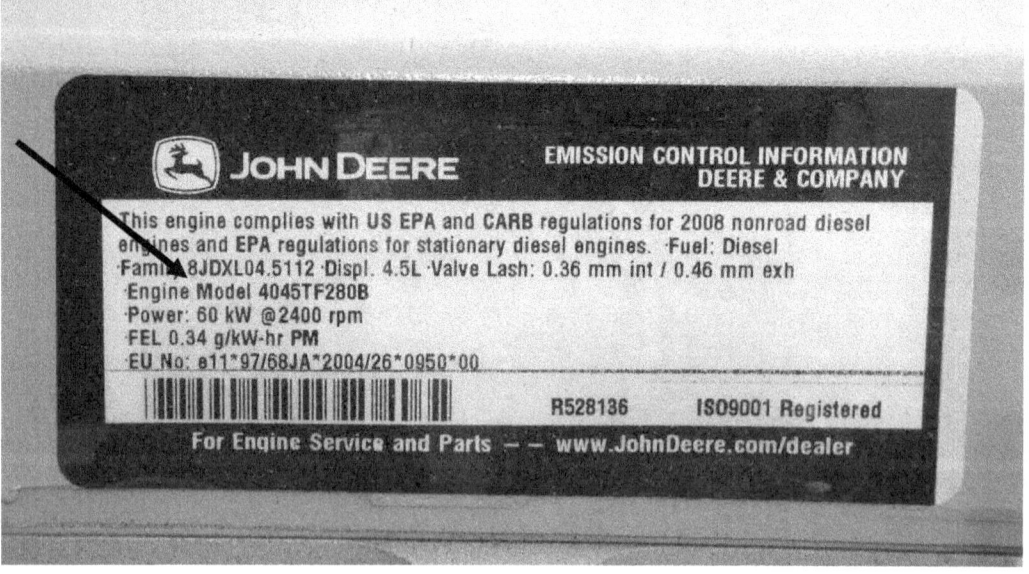

---

[4] http://www.tpub.com/content/altfuels10/epa/epa0005.htm
[5] http://www.epa.gov/OMS/cert/dearmfr/cd9107.pdf

Figure 4. Engine Family Label Inside the Engine Compartment (2)

Figure 5. Engine Family Label Inside the Engine Compartment (3)

Figure 6. Engine Family Label Inside the Engine Compartment (4)

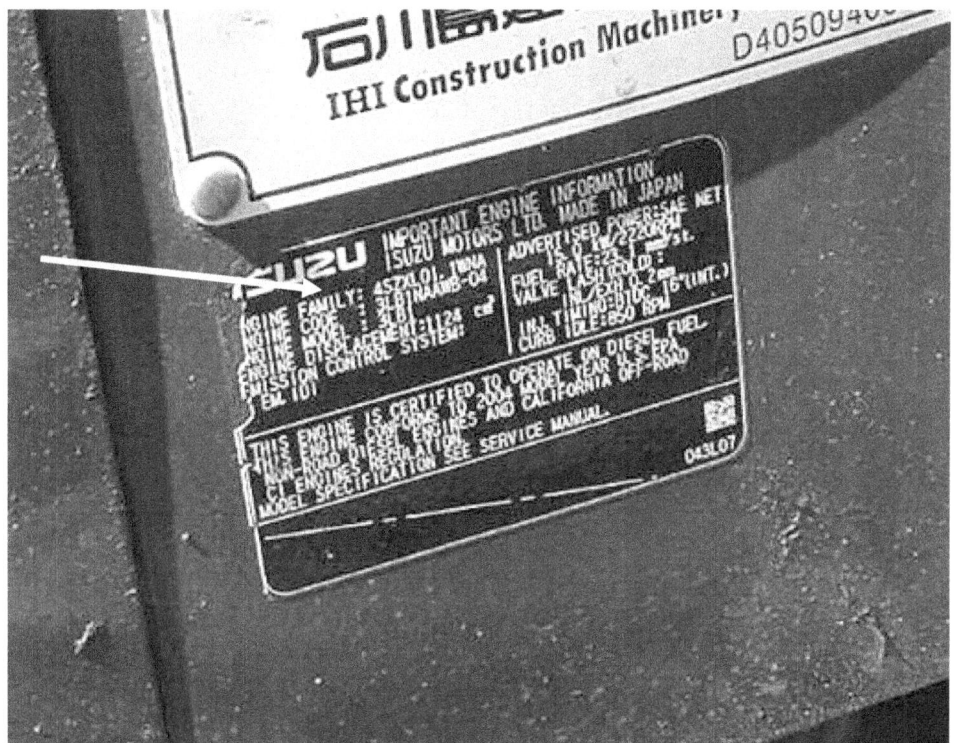

The engine family name provides the following information:
a. 1997 and earlier model years
    i. You can use Appendix F to decode the exhaust engine family code to obtain the following information:
        1. Model year
        2. Manufacturer
        3. Displacement
        4. Vehicle or engine class
        5. Fuel system and number of valves
        6. Combustion cycle and fuel
        7. Emission Standards
        8. Exhaust gas aftertreatment device (for example, a catalyst or a particulate trap)
        9. Whether on-board diagnostics regulations apply
    ii. The evaporative engine family code can be decoded according to the following:
        1. Character 1 – Model Year (this is the same as the exhaust engine family code)
        2. Character 2 & 3 – Manufacturer (this is the same as the exhaust engine family code)
        3. Character 4 – Vapor Storage System
            a. 1 = Canister
            b. 2 = Crankcase

   c. 3 = Air Cleaner

   d. 4 = Canister & Crankcase

   e. 5 = Crankcase & Air Cleaner

   f. 6 = Canister & Crankcase & Air Cleaner

  4. Characters 5, 6, & 7 – Canister work capacity (Total Grams in All Canisters)

  5. Character 8 – Canister Configuration

   a. W = Plastic Housing – Closed Bottom

   b. X = Plastic Housing – Open Bottom

   c. Y = Metal Housing – Closed Bottom

   d. Z = Metal Housing – Open Bottom

  6. Character 9 - Fuel System (this is the same as the exhaust engine family code)

  7. Character 10 – Fuel Tank Material

   a. M = Metal

   b. P = Plastic

  8. Character 11 – Purge[6] Control

   a. 1 = Controlled

   b. 0 = Not Controlled

  9. Character 12 – Wildcard (assigned by the manufacturer, but does not convey relevant information for building your inventory)

 iii. Example

The Class 8b heavy-duty diesel truck shown in Figure 7 has an engine family name of TDD12.EJDAR, as shown in Figure 8.

---

[6] This refers to the method in which the canister is purged.

Figure 7. Class 8b Heavy-duty Diesel Truck

Figure 8. Engine Family Name Label

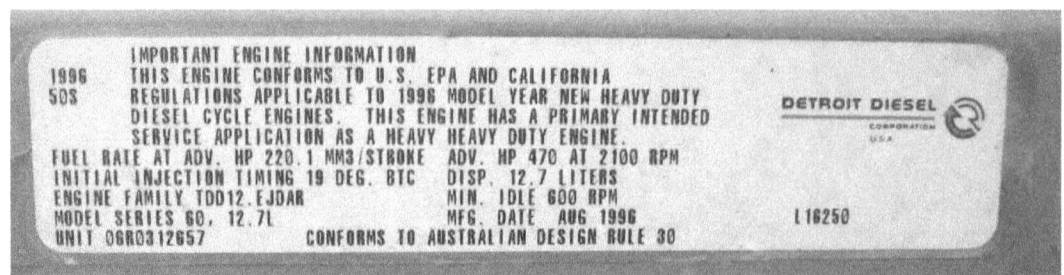

This engine family name can be decoded as:

T	1996
DD	Detroit Diesel
12.	12.7 Liter engine
E	Heavy -duty diesel / >14,000 GVW / HP >250
J	Electric Multi-Point Injection (MPI) - simultaneous / 3 or more Valves/Cylinders
D	Diesel Cycle (CI) / Diesel Fuel
A	Certified to Tier 0 emissions standards
R	Three-way + Oxidation catalyst
	Complies with federal OBD or California OBD II requirements

b. 1998 and later model years
   i. You can use Appendix F to decode the exhaust engine family code to obtain the following information:
      1. Model year
      2. Manufacturer
      3. Family type (a.k.a. vehicle or engine class)
      4. Displacement
      5. Sequence characters for family name
   ii. The evaporative engine family code can be decoded according to the following:
      1. Character 1 through 5 – These are the same as the exhaust engine family code
      2. Character 6, 7, & 8 – Canister Work Capacity (Total Grams in All Canisters)
      3. Characters 10, 11, & 12 – Sequence characters (This is a unique code assigned by the manufacturer to identify the engine family, but does not convey relevant information for building your inventory.)

iii. Example
The engine family name 3DDXH12.7EGY can be decoded as:

3	2003
DDX	Detroit Diesel
H	Heavy-duty engine family
12.7	12.7 liters
EGY	Manufacturer's code

Figure 9. Engine Family Label for Off-road Diesel Equipment (Example 1)

Figure 10. Engine Family Label for Off-road Diesel Equipment (Example 2)

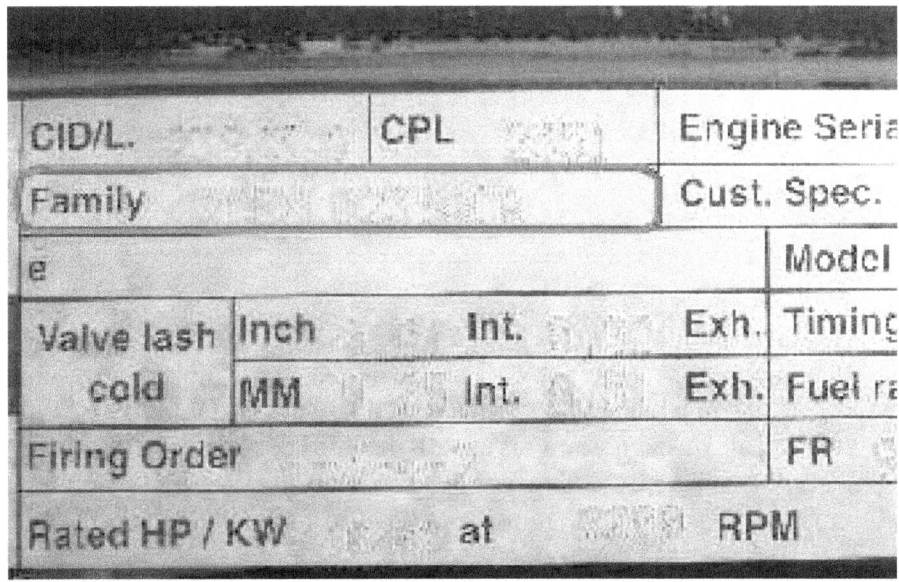

Figure 11. Engine Family Label for Off-road Diesel Equipment (Example 3)

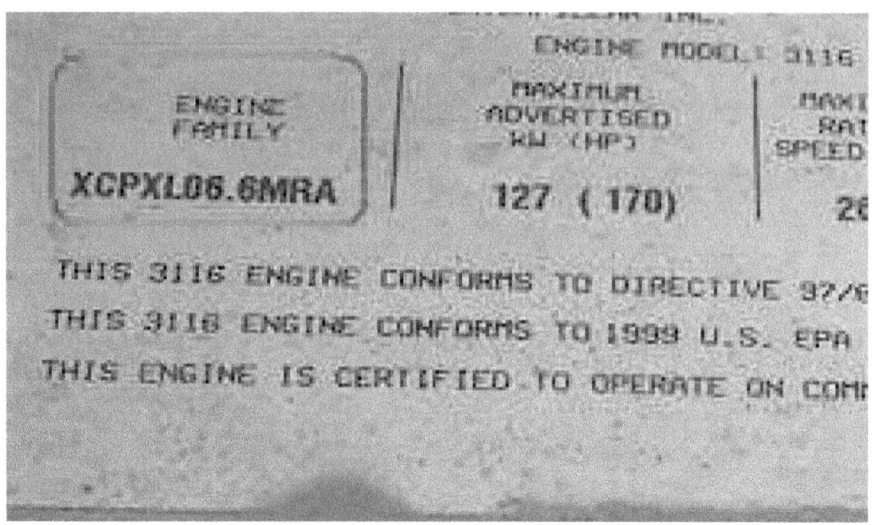

Figure 12. Engine Family Label for Off-road Diesel Equipment (Example 4)

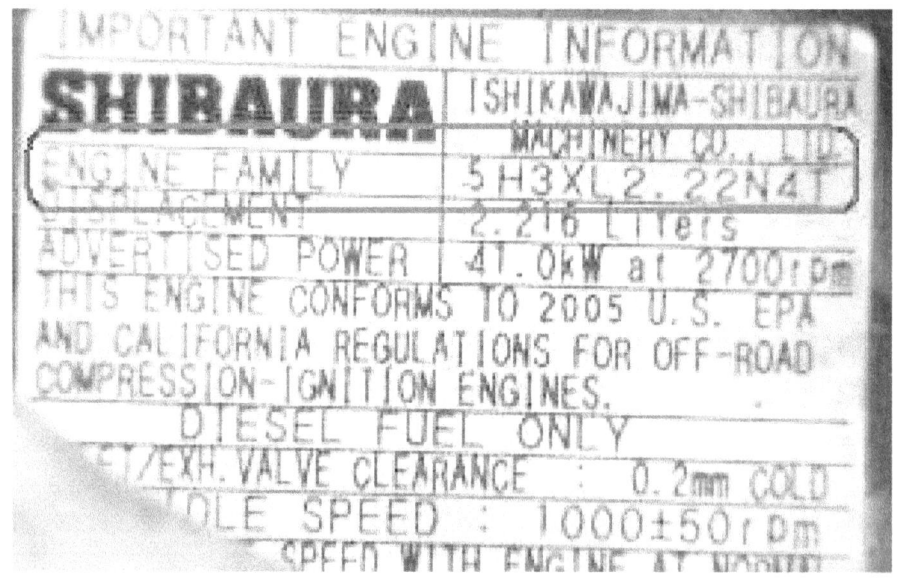

9.   **What are the make, model, and model year of the <u>engine</u>?**
As stated above, the engine family label will identify the engine make, model, and model year. The engine serial number can also help you identify this information about the engine, though you will usually have to contact the dealer or use their Internet resources to help identify this information from the serial number. Often the serial number and the engine family code are located near each other on the engine. Appendix H provides common locations of the serial numbers for both engines and chassis.

10. **What is the displacement and power rating for this engine?**

If you are unable to determine the displacement (in cubic centimeters or cubic inches) and power (in horsepower or kW)[7, 8] from the engine family name, you can use the engine and/or chassis make and model to look up information from the manufacturer's website or one of the many helpful websites on the Internet. Some examples of websites are mentioned in the "Useful Resources" section later in this guide. Appendix A also provides the contact information for many manufacturers.

11. **What type of fuel does this vehicle use?**

Fuel purchase records should identify what type of fuel was purchased for this vehicle/equipment. Otherwise, fuel information can be found on the engine family name label.

12. **What retrofits are used, if any?**

Control equipment or systems that are part of the original equipment are not retrofits and will be labeled on the engine family label. However, if an engine has undergone an after-market alteration or retrofit, you will likely have to report that information as part of the application process. Consult your purchase and maintenance records and your maintenance provider to determine if any after-market improvements have been made to the original equipment. This information will be very important when establishing the baseline emissions for your fleet. The information you may need regarding retrofits, if present include:

1. Retrofit type
2. Manufacturer
3. Model Name/Number
4. Year of installation
5. Size/Capacity

13. **What is the tier level of the engine?**

Off-road engines are certified to specific emission standards, referred to as the engine's "Tier Level". The engine horsepower rating and year of manufacture determine the tier level. Each tier level is phased-in over the span of several years according to horsepower rating. The engine tier level should be identified on the engine label (usually along with the engine family code). However, if you are unable to locate the appropriate label with the specific information for your engine, the table provided below can be used as a general guideline to help you identify which tier level your engine may be, *although this information is not definitive*. If you are unable to locate the engine label with the tier level certification clearly identified, it is recommended that you contact the engine or equipment manufacturer with the make, model, and model year information to obtain the correct tier level certification.

---

[7] Kilowatts can be converted into horsepower by multiplying by 1.34. [kW X 1.341 = HP].
http://www.arb.ca.gov/portable/perp/fleetemissions/calculatorinstructions.htm
[8] Most off-road engines report power in terms of horsepower. However, some equipment, such as generators, report power in terms of kilowatts (kW).

General Guide to EPA Tier Levels for Off-road Diesel Engines
by Horsepower Rating

Engine Power	Tier	Years	Engine Power	Tier	Years
hp < 11	Tier 1	2000 - 2004	175 ≤ hp < 300	Tier 1	1996 - 2002
	Tier 2	2005 - 2007		Tier 2	2003 - 2005
	Tier 4	2008 +		Tier 3	2006 - 2010
11 ≤ hp < 25	Tier 1	2000 - 2004		Tier 4	2011 +
	Tier 2	2005 - 2007	300 ≤ hp < 600	Tier 1	1996 - 2000
	Tier 4	2008 +		Tier 2	2001 - 2005
25 ≤ hp < 50	Tier 1	1999 - 2003		Tier 3	2006 - 2010
	Tier 2	2004 - 2007		Tier 4	2011 +
	Tier 4	2008 +	600 ≤ hp < 750	Tier 1	1996 – 2001
50 ≤ hp < 75	Tier 1	1998 – 2003		Tier 2	2002 – 2005
	Tier 2	2004 – 2007		Tier 3	2006 - 2010
	Tier 3	2008 +		Tier 4	2011 +
75 ≤ hp < 100	Tier 1	1998 – 2003	hp ≥ 750	Tier 1	2000 - 2005
	Tier 2	2004 – 2007		Tier 2	2006 - 2010
	Tier 3	2008[9]		Tier 4	2011 +
	Tier 4	2008 +			
100 ≤ hp < 175	Tier 1	1997 – 2002			
	Tier 2	2003 – 2006			
	Tier 3	2007 - 2011			
	Tier 4	2012 +			

14.   Is the engine turbocharged or naturally aspirated?

A naturally-aspirated (carbureted) engine depends on atmospheric pressure to draw in air for combustion. A turbocharged engine uses a blower to force additional air in for combustion, increasing the volume of air intake beyond what could be provided by atmospheric pressure alone. Most modern diesel engines are turbocharged, as turbocharged engines have higher power output and lower emission levels. You may be able to see the housing of the blower on the engine to tell if it is turbocharged or not, or you may be able to find that information on the engine label. Otherwise, you may need to consult your maintenance provider or the manufacturer.

15.   What type of lube oil is used?

You should find what type of oil is used from your maintenance records. Otherwise, consult with your maintenance provider to determine what type of oil is used in this vehicle/equipment. You may also be asked to provide lube oil consumption rates, which you should also be able to find from your maintenance records. Lube oil consumption rates will help determine if a given vehicle/equipment is eligible for certain retrofit technologies.

---

[9] 2008 is a Phase-in year for new engine standards. This is why a single year spans two different Tier levels. In these cases, refer to the engine label for clarification.

Helpful Hints

By far the easiest way to collect much of this information will be to check with your dealership, your maintenance providers, and your purchase records. However, if you need to collect the information directly from the equipment, the information you will be seeking is not always easy to see or to read. To help you see the engine labels and serial numbers, you may need a telescoping mechanic's mirror, as seen in Figure 13. These mirrors come in a variety of styles and are readily available at most auto parts stores or online.

Figure 13. Telescoping Mechanic's Mirror

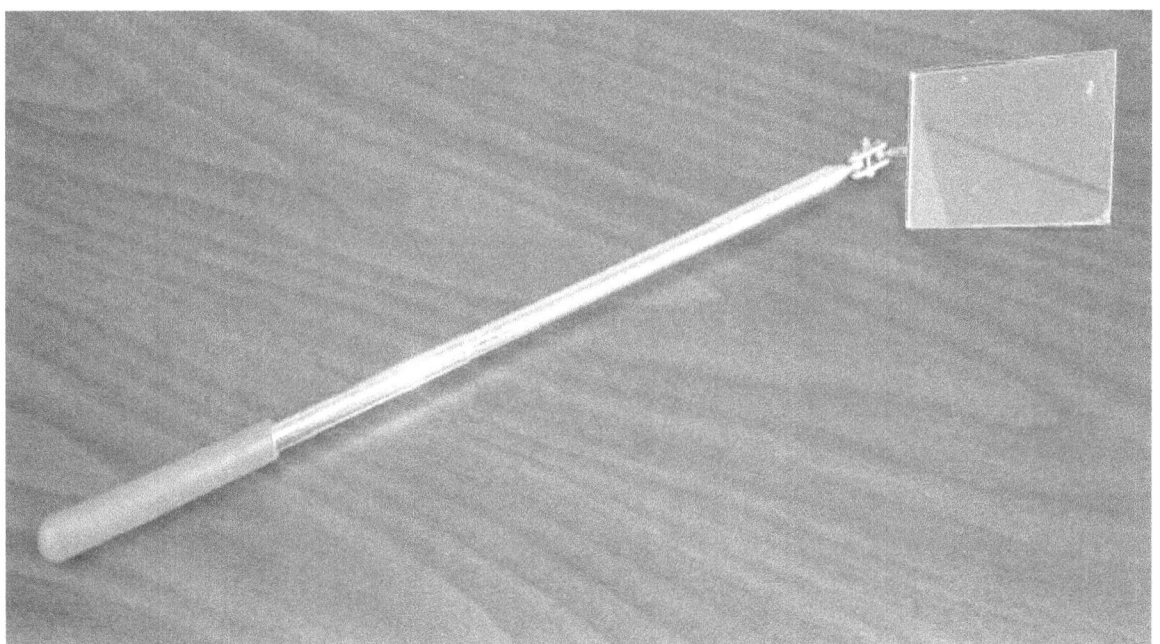

Also, because this equipment is used in construction, label and identification plates may become extremely dirty, especially inside the engine compartment. When you go to collect the information, be prepared to clean the labels so that you can read them. You may use a soft-bristle brush and some water to help remove the dirt from the label. If looking inside the engine compartment, have a solvent (such as WD-40) handy to help dissolve some of the oil and grime on the tags and labels. Sometimes, the information you are looking for has been stamped into a plate. Over time, the stamping can wear and become difficult to read. In these instances, you may find it helpful to take a rubbing of the plate using a pencil or crayon and a piece of paper.

Useful Resources

Each manufacturer has its own website with information for their specific equipment. Contact information for many common manufacturers is listed in Appendix A. Other helpful (free) resources for finding information such as engine displacement and horsepower include, but are not limited to the following websites:

- Equipment World Spec Guide
  - http://digitalmagazinetechnology.com/a/?KEY=equipmentworld-10-specguide#page=0
- Iron Record (for Caterpillar Equipment)
  - http://www.ironrecord.com/
- Spec-Check
  - http://www.spec-check.com/quickfinder_freetrial.cfm
- Equipment Watch
  - http://www.equipmentwatch.com/
- EPA Engine Certification Data
  - http://www.epa.gov/otaq/certdata.htm#largeng
- California Air Resources Board Off-road Certification Database
  - http://www.arb.ca.gov/msprog/offroad/cert/cert.php

Appendix A
Manufacturer Contact Directory

ABCO Engineering Corp.	ASV Inc.	Aeroil Products Co., Inc.
Oelwein, IA	Grand Rapids, MN	Crossville, TN
319-283-5652	218-327-3434	931-456-8855
www.abcoveyor.com	www.asvi.com	
Aerolift	Akerman	Alitec
Caldwell, NJ	(see Volvo)	Woods Equipment Company
973-575-7484		Oregon, IL
www.aero-lift.com		800-848-3447
		www.alitec.org
Allen Engineering	Allied Construction Products, Inc.	Allied Systems Company
Paragould, AR	Cleveland, OH	Sherwood, OR
800-643-0095	216-431-2600	503-625-2560
www.alleneng.com	www.alliedcp.com	www.alliedsystems.com
Allmand Bros., Inc.	Ameriquip	American Augers, Inc.
Holdrege, NE	Kiel, WI	West Salem, OH
308-995-4495	920-894-7063	419-869-7107
www.allmand.com	www.amerequip.com	www.americanaugers.com
American Crane Corp.	American Jenbach Corp.	American Piledriving Equipment,
A Terex Cranes Division Business Unit	(See Burco Welding	Inc.
Wilmington, NC		Kent, WA
910-395-8500		253-872-0141
www.terex-cranes.com		www.apevibro.com
American Pneumatic Tool, Inc.	Ammann American inc.	ARPS
Gardena, CA	Farmers Branch, TX	(See Amerequip)
310-538-2600	972-488-2233	
www.apt-tools.com	www.ammann-america.com	
ARDCO/Traverse Lift LLC	Armlift	Arrow Manufacturing Co.
(See Pettibone)	Division of TG Industries, Inc.	(See Arrow-Master, Inc.)
	Armstrong, IA	
	712-864-3737	
	www.armlift.com	
Arrow-Master, Inc.	Astec Underground	ATECO American Tractor
East Moline, IL	Loudon, TX	Equipment Corp.
309-752-1345	865-408-2100	Logan, UT
www.arrowmaster.com	www.astecunderground.com	435-755-9383
Athey Products Corp.	Atlas Copco Compressors, Inc.	Badger Construction Equipment
(See Elgin Sweeper Co.)	Westfield, MA	Co.
	413-536-0600	Winona, MN
	www.atlascopco.com	507-454-1563
		www.badgerequipment.com
Balderson, Inc.	Bantam Division	Baraga products (BPI)
(See Caterpillar)	(See Koehring Co. – Bantam	(See Terex)
	Products)	
Barger-Greene Div.	Bartell Industries	Bell Equipment
(See Caterpillar Paving Products)	(See Terex Bartell Ltd.)	Garden City, GA 31408
		912-966-2615
		www.bell.co.za
Benford	Beuthling Mfg. Co.	Bid-Well
(See Terex Light Construction)	Clear Lake, WI	(See Terex Roadbuilding)
	715-263-2300	
	www.beuthling.com	

Blaw Knox (See Ingersoll Rand Construction Technologies, Road Development Division)	Blount, Inc. Zebulon, NC 27597 919-269-7421 www.blount-fied.com	Bobcat Company West Fargo, ND 701-241-8700 www.bobcat.com
Bomag Americas, Inc. Kewanee, IL 309-853-3571 www.bomag.com	Boss Industries, Inc. LaPorte, IN 219-324-7776 www.bossair.com	Braden-Carco-Gearmatic PACCAR WINCH DIVISION Broken Arrow, OK 918-251-8511 www.paccarwinch.com
Broderson Manufacturing Corp. Lenexa, KS 913-888-0606 www.bmccranes.com	Bucyrus International, Inc. South Milwaukee, WI 414-768-4000 www.bucyrus.com	Buffalo-Bomag & Buffalo-Springfield (See Bomag)
Burco Welding & Cutting Products, Inc. High Point, NC 336-887-6100 www.burco.net	Burkeen Mft. Co. Olive Branch, MS 662-895-4150 www.burkeen.com	C H & E Manufacturing Co. ABS Dewatering Division Milwaukee, WI 800-236-0666
CMI Corp. (CMI Terex) (See Terex Roadbuilding)	CMI Environmental Machinery (See Terex Roadbuilding)	CMI Johnson-Ross (See Terex Roadbuilding)
Calavar Corporation (See Time Condor)	Calder Brothers Corporation Greenville, SC 864-244-4800	Campbell International, Inc. Wauconda, IL 847-526-7300 www.campbellcab.com
Carco Winch Products (See branden-Carco-Gearmatic)	Case Construction Equipment Racine, WI 262-636-6011 www.casece.com	Caterpillar Attachment Products & Services Wamego, KS 800-255-2372 www.cat.com
Caterpillar, Inc. Peoria, IL 309-675-1000 www.cat.com	Caterpillar Paving Products Minneapolis, MN 612-493-1317 www.cat.com	Cedarrapids Inc. A Terex Roadbuilding Co. Products Group Business Unit Cedar Rapids, IA 319-363-3511 www.cedarapids.com
Century II (See Terex Cranes)	Champion Road Machinery (See Volvo)	Chicago Pneumatic Tool Co. Atlast Copco Construction Tools West Springfield, MA 800-760-4049 www.chicagopneumatic.com
CIFA USA Yorkville, WI 262-835-1825 www.cifausa.com	Cleveland Trencher Akron, OH 330-869-2800 www.cleveland-trencher.com	CMI Terex (See Terex Roadbuilding)
Compact Technologies (See Volvo Construction Equipment)	CompAir America Sidney, OH 937-498-2500 www.compair.com	Concrete Equipment Co. Blair, NE 402-426-4181 www.con-e-co.com
Concrete Surfacing Machinery Division of Stewart Industries, Inc. Cincinnati, OH 513-891-9000 www.stewartindustries.com	Condor (See Time Condor Corporation)	Contractors Manufacturing Services Inc. (See Robbins HDD)

Converto Manufacturing Cambridge City, IN 765-478-3205 www.convertomfg.com	Coyote Equipment Hudson, OH 330-650-5101 www.coyoteloaders.com	Crane Carrier Company Tulsa, OK 918-836-1651 www.cranecarrier.com
Crisafulli Pump Co., Inc. (SRS Crisafulli) Glendive, MT 406-365-3393 www.crisafulli.com	Cummins Inc. Columbus, IN 812-377-5000 www.cummins.com	Curbmaster USA, Inc. (See CMI)
Daewoo Doosan Infracore America Corp Suwanee. GA 770-831-2200 www.usa.doosaninfracore.co.kr	Dart Truck Co. (See Unit Rig)	Davey Drill (See Davey Kent)
Davey Kent Kent, OH 330-673-5400 www.daveykent.com	Davis Manufacturing (See Case Corp.)	Deere & Company Moline, IL 309-765-8000 www.deere.com
Demag Cranes & Components A Terex Compay Wilmington, NC 910-395-8500 www.terex-cranes.com	Desa International Bowling Green, KY 270-781-9600 www.desaint.com	Detroit Diesel Corp. A Daimler Chrysler Company Detroit, MI 313-592-5000 www.detroitdiesel.com
Digmor California Fontana, CA 909-381-0333	Ditch Witch The Charles Machine Works, Inc. Perry, OK 580-336-4402 www.ditchwitch.com	Dressta North America Ltd. Buffalo Grove, IL 847-537-4783 www.dressthnorthamerica.com
Driltech Mission, LLC A Sandvik Company Alachua, FL 386-462-4100 www.driltech.sandvik.com	Drott (See Case Corp.)	Duratech Industries Jamestown, ND 701-252-4601 www.dura-ind.com
Dynapac Compaction, Paving & Milling Schertz, TX 210-474-5770 www.dynapac.com	E.D. Etnyre & Company Oregon, IL 815-732-2116 www.etnyre.com	Eager Beaver Lake Wales, FL 863-638-1421 www.eagerbeavertrailers.com
Eagle Iron Works Des Moines, IA 515-243-1123 www.eagleironworks.com	Elco International Inc. Elmwood park, NJ 201-797-4644 www.elco.com	Electric Tower Cranes, Inc. (See Elco International, Inc.)
Elgin Sweeper Co. Elgin, IL 847-741-5370 www.elginsweeper.com	El-Jay, Inc. (See Cedarapids)	Elkin Manufacturing Inc. Indiana, PA 724-349-6300 www.elkinhitech.com
Elliott Equipment Company Omaha, NE 402-592-4500 www.elliottequip.com	Emaco/Elco Int'l. (See Elco International Inc.)	Erie Strayer Company Erie, PA 814-456-7001 www.eriestrayer.com
Esco Corporation Portland, OR 503-228-2141 www.escocorp.com	Essick Manufacturing Co. (See Multi-Quip, Inc.0	Euclid-Hitachi (See Hitachi)

Fairbanks Morse Pump Corp. Kansas City, KS 913-371-5000 www.fairbanksmorsepump.com	Federal Signal Corp. Oak Brook, IL 630-954-2000 www.federalsignal.com	Ferguson Manufacturing & Equipment Co. Dallas, TX 214-631-3000
Fermec North America (See Terex Construction Americas)	Ferree Trailer/Vision Metals, Inc. Liberty, NC 336-622-7300 www.ferreetrailers.com	Finlay USA A Terex Company Louisville, KY 502-736-5260 www.finlayhydrascreens.com
Flow Boy Mfg. A Hi-Way Company Norman, OK 405-329-3765 www.flowboy.com	FMC Corporation (See Johnston)	Fontaine Trailer Company Haleyville, AL 800-821-6535 www.fontainetrailer.com
Ford Motor Co. Ford Truck Operations Dearborn, MI 313-328-9707 www.ford.com	Ford New Holland (See New Holland)	Franklin Treefarmer Franklin, VA 757-562-6111 www.franklin-treefarmer.com
Freuhauf Trailer Corp. (See Wabash National)	Galion Manufacturing Division (See Komatsu)	Gar-Bro Manufacturing Company Herber Springs, AR 501-362-8171 www.garbro.com
Gardner Denver, Inc. Blower Division Peachtree City, GA 770-632-5000 www.gardnerdenver.com	Gardner Denver, inc Compressors Quincy, IL 217-222-5400 www.gardnerdenver.com	Gardner Denver, Inc. Pump Division Tulsa, OK 918-664-1151 www.gardnerdenver.com
Gehl Co. West Bend, WI 262-334-9461 www.gehl.com	Gencor Industries, Inc. Orlando, Fl 407-290-6000 www.gencor.com	Generac Corp. Waukesha, WI 888-436-3722 www.generac.com
Genie North America A Terex Company Redmond, WA 425-881-1800 www.genielift.com	Gomaco Corporation Ida Grove, IA 712-364-3347 www.gomaco.com	The Gorman-Rupp Company Mansfield, OH 419-755-1011 www.gormanrupp.com
Gradall Industries, inc. New Philadelphia, OH 330-339-2211 www.gradall.com	Gradall Telehandler (See JLG Industries, Inc.)	Grasan Equipment Co., Inc. Mansfield, OH 419-526-4440 www.grasan.com
Grimmer-Schmidt Compressor Franklin, IN 317-736-8416 www.grimmerschmidt.com	Griswold Machine & Engineering Union City, MI 517-741-4471 www.gmeco.com	Grove Worldwide A Subsidiary of the Manitowoc Company, Inc. Shady Grove, PA 717-597-8121 www.manitowoccranes.com
Greundler/Simplicity Crusher A Terex Company Durand, MI 989-288-3121 www.greundlercrusher.com	Guest Industries Torrington, CT 860-482-1118 www.guestindustries.com	Guntert & Zimmerman Const. Div., Inc. Ripon, CA 209-599-0066 www.guntert.com

H-R Mixer Corporation (See T.L. Smith Machine Co.)	H & S Co., Inc. Celina, OH 419-394-4444 www.wheeledtrenchers.com	Hahn Machinery, Inc. Two Harbors, MN 218-834-2156 www.hahnmachinery.com
Hamm Compaction Division Wirtgen America, Inc. Nashville, TN 615-501-0600 www.hammcompactors.com	Harnischfeger Corporation (P&H) (See Terex Cranes)	Hendrix Manufacturing Company, Inc. Mansfield, LA 318-872-1660 www.hendrixmfg.com
Hewitt-Robins Pueblo West, CO 800-388-7701 www.hewitt-robins.com	Hitachi Construction Machinery Co. A Deere Company Affilliate Moline, IL 309-765-8000 www.hitachiconstruction.com	Hobart Brothers Company Troy, OH 937-332-4000 www.hobartbrothers.com
Homelite Consumer Products, Inc. Charlotte, NC 800-242-4672 www.homelite.com	Huber/Scott (See Multi-Quip, Inc.)	Hy-Dynamic Division (See Koehring)
Hydro-Ax (See Blount, Inc.)	Hypac Bomag Americas, Inc. Kewanee, IL 309-853-3571 www.hypac.com	Hyster (See Hypac)
Hyundai Construction Elk Grove, IL 847-437-3333 www.hceusa.com	IHI (Ishikawajima-Harima Heavy Industries) Co., Ltd. Compact Excavator Sales, LLC Elizabethtown, KY 800-538-1447 www.ihicompactexcavator.com	IMT (Iowa Mold Tooling) Co., Inc. Des Moines, IA 800-717-1177 www.imt.com
Ingersoll Rand Company Montavale, NJ 201-573-0123 www.ingersollrand.com	Ingersoll Rand Construction Technologies Road Development Division Shippensburg, PA 717-532-9181 www.road-development.irco.com	Ingram Manufacturing Company (See Pavement Services Inc.)
Insley Manufacturing Corp. Winona, MN 507-454-1563	International-Hough Div. Dresser Industries (See Komatsu)	JCB, Inc. Pooler, GA 912-447-2000 www.jcb.com
JCB Vibromax www.jcbvibromax.de	JLG Industries, Inc. Mcconnellsburg, PA 717-485-5161 www.jlg.com	JRB Co., Inc. Akron, OH 330-734-3000 www.jrbco.com
The Jay Company (See Davey Kent)	C.S. Johnson Co. (See CMI Johnson-Ross Corp.)	Johnston Sweeper Co. www.johnstonsweepers.com
Joy Manufacturing Company (See Sullivan Machinery Co.)	K-D Manitou, Inc. (See Manitou North America)	Kato (See Mitsui)
Kawasaki Construction Machinery Corp. of America Kennesaw, GA 770-499-7000 www.kawasakiloaders.com	Kent Demolition Tool Co. Kent, OH 800-527-2282 www.kentdemolition.com	Klein Products, Inc. Ontario, CA 909-460-4546 www.kleinproducts.com

Kobelco Construction Machinery America LLC Carol Stream, IL 866-726-3396 www.kobelcoamerica.com	Koehring Company (See Terex Cranes)	Kohler Co., Engine Division Kohler, WI 920-457-4441 www.kohlerengines.com
Kolbert-Pioneer, Inc. An Astec Industries Company Yankton, SD 605-665-8771 www.kolbergpioneer.com	Kolman Sibley, IA 712-754-4661 www.kolman.com	Komatsu America International Co. Vernon Hills, IL 847-970-4100 www.komatsuamerica.com
Komatsu Forest LLC Shawano, WI 715-524-2820 www.komatsuforest.com	Krause Manufacturing Co. (See Simon Aerials)	Kubota Tractor Corp. Torrance, CA 310-370-3370 www.kubota.com
LBX Company, LLC Lexington, KY 859-245-3900 www.lbxco.com	LDC Industries, Inc. (See Arrow-Master, Inc.)	Lay-Mor Longfivew, TX 800-323-0135 www.laymor.com
Layton Manufacturing Co., Inc. Salem, OR 503-585-4888	LeRoi International, Inc. (See CompAir)	LeTourneau Inc., Equipment Group Longview, TX 903-237-7000 www.letourneau-inc.com
LeeBoy Denver, NC 704-966-3300 www.leeboy.com	Liebherr-America, Inc. Newport News, VA 757-245-5251 www.liebherr.com	Lift-A-Loft Muncie, IN 765-288-3691 www.liftaloft.com
Liftking Industries, Inc. Woodbridge, ON Canada 905-851-3988 www.liftking.com	The Lincoln Electric Company Cleveland, OH 216-481-8100 www.lincolnelectric.com	Link-Belt Construction Equipment Company Lexington, KY 859-263-5200 www.linkbelt.com
Link-Belt Earthmoving (See LBX Company LLC)	Lister-Petter, Inc. Olathe, KS 913-764-3515 www.lister-petter.com	Little Giant Pump Co. Oklahoma City, OK 405-947-2511 www.lgpc.com
Load King (See Terex Load King)	Lombardini USA, Inc. Duluth, GA 770-623-3554 www.lombardiniusa.com	Long Agribusiness LLC Tarboro, NC 252-823-4151 www.farmtrac.com
Lorain (See Terex Cranes)	Lull International, Inc. (See JLG Industries, Inc.)	M-B-W, Inc. Slinger, WI 262-644-5234 www.mbw.com
MCE (See Mitsui)	MF Industrial Machinery (See Terex Construction Americas)	MKT Manufacturing, Inc. St. Louis, MO 314-388-2254 www.mktpileman.com
MMD Equipment Swedesboro, NJ 856-467-3200 www.mmdequipment.com	MP Pumps, Inc. Fraser, MI 586-293-8240 www.mppumps.com	Mack Trucks, Inc. – World Headquarters Allentown, PA 610-709-3011 www.macktrucks.com

Manitex Georgetown, TX 512-942-3000 www.manitex.com	Manitou North America Waco, TX 254-799-0232 www.manitou-na.com	Manitowoc Crane Group Manitowoc, WI 920-684-6621 www.manitowoccranes.com
Mannesmann Dematic Corp. (See Demag cranes & Components)	Mantis Cranes (See SpanDeck, Inc.0	E.F. Marsh Engineering Co. St. Louis, MO 314-968-4700
Master Craft Industrial Equipment Tifton, GA 229-386-0610 www.mclifts.com	Mauldin Paving Products (See Calder Brothers Corporation)	Maxon Industries, Inc. Milwaukee, WI 414-351-4000 www.maxon.com
Mayco Pump Corp. (See Multiquip)	Mayville Engineering Company, Inc. (MEC) Mayville, WI 920-387-4500 www.mayvl.com	Messinger, Inc. Salisbury, NC 704-638-0405 www.messingerinc.net
Mikasa Construction Equipment Div. (See Multiquip, Inc.)	Miller Spreader Co. Youngston, OH 800-377-4565 www.millerspreader.com	Mobile Sweeper Div. (See Elgin Sweeper Co.)
Morgen Manufacturing Company MinnPar Inc. Minneapolis, MN 612-379-0606	Moxy Trucks Cincinnati, OH 513-831-2000 www.moxytrucks.com	Muller Machinery Co., Inc. (See Terex Light Construction)
Multiquip, Inc. Carson, CA 310-537-3700 www.multiquip.com	Mustang Manufacturing Co., Inc. A Gehl Company Owatonna, MN 507-451-7112 www.mustangmfg.com	NPK Construction Equipment Walton Hills, OH 440-232-7900 www.npkce.com
National Crane Corporation Mantiowoc Crane Gorup Shady Grove, PA 717-597-8121 www.manitowoccranes.com	Newstripe Inc. Aurora, CO 303-364-7786 www.newstripe.com	New Holland Construction Carol Stream, IL 630-260-4000 www.newhollandconstruction.com
Niftylift, Inc.-USA Glen Ellyn, IL 630-858-0822 www.niftylift.com	Northwest Engineering Company MinnPar Inc. Minneapolis, MN 612-379-0606 www.minnpar.com	O & K (Orenstein & Koppel) Inc. Terex Mining Tulsa, OK 918-446-5581 www.ok-mining.com
Okada America Clackamas, OR 503-557-7033 www.okadaamerica.com	Onan Corp. (See Cummins Engine Co., Inc.)	Owatonna Manufacturing Co., Inc. (See Mustang Mfg.)
PPM Cranes/P&H (See Terex Cranes)	Pacific Car & Foundry Co. (See Braden-Carco-Gearmatic)	Parsons Trenchers (See Maxon Industries, Inc.)
Partek Forest LLC (See Komatsu Forest LLC)	Pavement Services Inc. (Ingram) Madison, SD 605-256-0795 www.askpsi.com	Payhauler Corp. (See Caterpillar, Inc.)
Peerless Conveyor and Manufacturing Corp. Kansas City, KS 913-342-2240 www.peerlessconveyor.com	Perkins Engines, inc. Mossville, IL 309-578-7364 www.perkins.com	Pettibone Baraga, MI 906-353-6611 www.pettiboneusa.com

Pioneer (See Kolberg-Pioneer, Inc.)	Poclain Div. – Case (See Case)	Power Curbers, Inc. Salisbury, NC 704-636-5871 www.powercurbers.com
Puckett Mft., Inc. Loganville, GA 877-218-3240 www.puckettmfg.com	Quincy Compressor Quincy, IL 217-277-0343 www,quincycompressor.com	R O Corporation (See Terex Cranes)
Rammax (See Multiquip, Inc.)	Rammer Inc. A Sandvik Tamrock USA Co. Cleveland, OH 216-431-2600 www.rammer.sandvik.com	Ranco Trailers Lamar, CO 719-336-9041 www.rancotrailers.com
Ranger (See Allied Systems Company)	Raygo, Inc. (See Catepillar Paving Products)	Read Corp., The Middleboro, MA 508-946-1200
Red River Mft. A Division of Trail King Industries, Inc. West Fargo, ND 701-282-3013 www.redrivermfg.com	Reedrill A Terex Company Sherman, TX 903-786-2981 www.reedrill.com	Rexworks/Rexnord (See CMI Environmental machinery)
Rice Hydro, Inc. Carson City, NV 775-885-1280 www.ricehydro.com	Rivinius/Domor Incorporated Eureka, IL 309-467-2303 www.rivinius-domor.com	Roadtec, Inc. Chattanooga, TX 423-265-0600 www.roadtec.com
Robbins HDD A Division of the Robbins Company Cleveland, OH 216-334-1000 www.robinshdd.com	Rogers Brothers Corporation Albion, PA 814-756-4121 www.rogerstrailers.com	Rosco Manufacturing co. (See Lee Boy)
Saf-T-Cab, Inc. Fresno, CA 559-268-5541 www.saftcab.com	Sakai America, Inc. Adairsville, GA 770-877-9433 www.sakaiamerica.com	Samsung Construction Equip. Co. (See Volvo)
Sandvik Tamrock LLC Sandvik Mining and Construction LLC Atlanta, GA 404-589-3800 www.smc.sandvik.com	Scat Trak (See Volvo Construction Americas)	Schaeff of North America, Inc. (See Terex Construction Americas)
Schramm, Inc. West Chester, PA 610-696-2500 www.schramminc.com	Schwarze Industries, inc. Huntsville, AL 256-851-1200 www.schwarze.com	Schwing America, Inc St. Paul, MN 651-429-0999 www.schwing.com
Scott Huber (See Multi-Quip, Inc.0	Shovel Supply Company, Inc. (See Ferguson Mfg.)	Shuttlelift, Inc. Sturgeon Bay, WI 920-743-8650 www.shuttlelift.com
Simon Aerials, Inc. (See Terex Lifting)	Sioux Steam Cleaner Corporation Beresford, SD 605-763-3333 www.sioux.com	Skyjack Guleph, Ontario, Canada 519-837-0888 www.skyjackinc.com

Sky Trak (See JLG Industries, Inc.)	Smith Air Compressors (See Boss Industries, Inc.)	Snorkel International St. Joseph, MO 785-989-3000 www.snorkleusa.com
SpanDeck, Inc. Franklin, TX 615-794-4556 www.mantiscranes.com	Stone Construction Equipment, Inc. Honeoye, NY 800-888-9926 www.stone-equip.com	Stow Manufacturing Co. (Multiquip) Carson, CA 310-661-4242 www.stomfg.com
StraightLine Manufacturing Newton, KS 800-654-3484 www.straightlinehdd.com	Strato-Lift International Corp. Parts for Lifts, Inc. North wales, PA 215-699-1701 www.stratolift.com	Sullair Corporation Michigan City, IN 219-879-5451 www.sullair.com
Sullivan-Palatek Portable Division Claremont, NH 603-543-3131 www.sullivanind.com	Superior Equipment Co. Buckeye Division (See H&S Co., Inc.)	SuperPac Compaction A Division of Volvo Construction Equipment Skyland, NC 828-684-3121 www.superpac.com
Svedala Dynapac (See Dynapac)	Svedala Reedrill (See Reedrill)	Sweepster, Inc. Paladin Light Construction Dexter, MI 800-456-7100 www.sweepster.com
TCI Power Products, Inc. (See Gehl)	TCM America MMD Equipment Swedensboro, NJ 800-433-1382 www.mmdequipment.com	Tadano America Corporation Houston, TX 281-869-0300 www.tadanoamerica.com
Takeuchi Mfg., (US), Ltd. Buford, GA 770-831-0661 www.takeuchi-us.com	Talbert Manufacturing, Inc. Rensselaer, IN 219-866-7141 www.talbertmfg.com	Tamrock (See Sandvik Tamrock LLC)
Target Products Olathe, KS 913-928-1000 www.targetblue.com	Taylor Machine Works, Inc. Louisville, MS 662-773-3421 www.taylorbigred.com	Terex Bartell Ltd. Barmpton, ON Canada 905-458-5455 www.terex.com
Terex Construction Americas Southaven, MS 662-393-1800 www.terexamericas.com	Terex Cranes, Inc. A Division of Terex Corp. Waverly, IA 319-352-3920 www.terex-cranes.com	Terex Cranes Wilmington Operations Wilmington, NC 910-395-8500
Terex Earthmoving-Crushing & Screening (See Cedarapids, Inc.)	Terex Earthmoving-Mining (See Caterpillar, Inc.)	Terex Lifting (See Terex Cranes)
Terex Light Construction Rock Hill, SC 803-324-3011 www.terex.com	Terex Load King Elk Point, SD 605-356-3301 www.loadkingtrailers.com	Terex Roadbuilding Oklahoma City, OK 405-787-6020 www.terexrb.com
Terex Utilities 888-837-3977 www.terexutilities.com	Terramite Corp. Cross Lanes, WV 304-776-4231 www.terramite.com	Tesmec USA, Inc. Alvarado, TX 817-473-2233 www.tesmec.com

Thomas Equipment Ltd. Centreville, NB Canada 506-276-4511 www.thomasloaders.com	Timbco (See Komatsu Forest LLC)	Timberjack Forestry Group A John Deere Company Moline, IL 309-765-1859 www.deere.com
Time Condor Corp. (See Stratolift)	Trail-Eze Trailers Mitchell, SD 605-996-6482 www.traileze.com	Trail King Mitchell, SD 605-996-6482 www.trailking.com
Tramac Corp. Parsippany, NJ 973-887-7700 www.tramac.com	Traverse Lift, LLC (See Pettibone)	Tree Farmer Equipment Co., Inc. (See Franklin Treefarmer)
Trencor Inc. As Astec Company Grapevile, TX 817-424-1968 www.trencor.com	Unit Rig (See Caterpillar, Inc.0	Valmet (See Komatsu Forest LLC)
Vermeer Manufacturing Co. Pella, IA 641-628-3141 www.vermeer.com	Vibro-Plus (See Dyna-Pac Mfg., Inc.)	Vibromax America, Inc. (See JCB Vibromax)
Volvo Construction Equipment Asheville, NC 828-650-2000 www.volvo.com	Volvo Motor Graders Goderich, ON Canada 519-524-2601 www.volv.com	Wabash National Lafayette, IN 765-771-5300 www.wabashnational.com
WABCO (See Komatsu)	Wacker Corp. Menomonee Falls, WI 262-255-0500 www.wachergroup.com	Waldon, Inc. Industrial Division Longview, TX 866-283-2759 Sweepmaster Division 800-323-0135 www.waldonequipment.com
Warner & Swasey Company Gradall Div. (See Gradall Industries, Inc.)	Warren Rupp Company Unit of IDEX Corp. Mansfield, OH 419-524-8388 www.warrenrupp.com	Watson, Inc. Fort Worth, TX 817-927-8486 www.watsonusa.com
Waukesha Engine Div. Waukesha, WI 262-547-3311 www.waukeshaengine.com	Whiteman Enterprises (See Multiquip, Inc.)	WINCO, Inc. LeCenter, MN 507-357-6821 www.wincogen.com
Wirtgen America Nashville, TN 615-501-0600 www.wirtgenamerica.com	Wood/Chuck Chipper Corp. Shelby, NC 800-269-5188 www.woodchuckchipper.com	Worthington Compressors, Inc. (See Atlas Copco)
Yamaha Motor Corp. USA Cypress, CA 800-962-7926 www.yamaha-motor.com		

Appendix B
On-Road Vehicle Weight Classifications

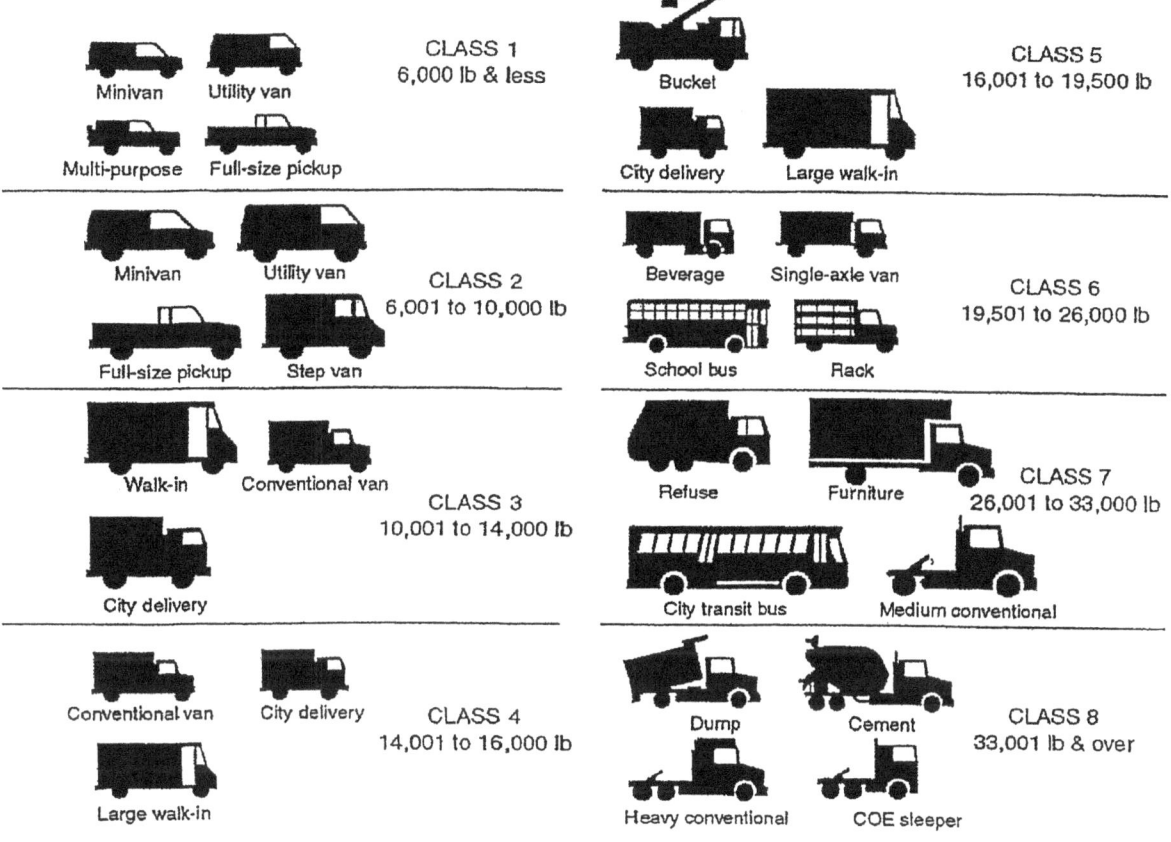

CLASS 1
6,000 lb & less
(Minivan, Utility van, Multi-purpose, Full-size pickup)

CLASS 2
6,001 to 10,000 lb
(Minivan, Utility van, Full-size pickup, Step van)

CLASS 3
10,001 to 14,000 lb
(Walk-in, Conventional van, City delivery)

CLASS 4
14,001 to 16,000 lb
(Conventional van, City delivery, Large walk-in)

CLASS 5
16,001 to 19,500 lb
(Bucket, City delivery, Large walk-in)

CLASS 6
19,501 to 26,000 lb
(Beverage, Single-axle van, School bus, Rack)

CLASS 7
26,001 to 33,000 lb
(Refuse, Furniture, City transit bus, Medium conventional)

CLASS 8
33,001 lb & over
(Dump, Cement, Heavy conventional, COE sleeper)

---

[10] Image developed by University of California, Riverside, Bourns College of Engineering, Center for Environmental Research and Technology.

Appendix C
Off-Road Equipment Definitions and Examples

Off-Road Equipment[11]

❖ **Loaders**
  ➢ Also called bucket loaders or front-end loaders
  ➢ Have a front-mounted bucket for scooping, though other attachments can be used instead of a bucket.
  ➢ Source Classification Code is 2270002060 for diesel-fueled loaders

Source: http://en.wikipedia.org/wiki/File:Wheel-loader02.jpg

---

[11] "User's Guide for the Final NONROAD2005 Model.", United States Environmental Protection Agency, December 2005.

❖ Pavers
- ➤ Large and small (such as for curbs)
- ➤ Primarily self-propelled pavers
- ➤ Source Classification Code is 2270002003 for diesel-fueled pavers

❖ Rollers/Compactors
- ➤ Includes smooth and knobby rollers
- ➤ Self-propelled rollers
- ➤ Source Classification Code is 2270002015 for diesel-fueled rollers

Source: Photo taken by Jan Mehlich, 29 September 2006, http://en.wikipedia.org/wiki/File:Dynapac_CC232.JPG

❖ **Scrapers**
  ➢ An off-highway tractor with a mid-frame bucket that lowers to scrape loose material (dirt) into the bucket to carry to another part of the job site to dump.
  ➢ Scrapers can be converted to water wagons.
  ➢ Source Classification Code is 2270002018 for diesel-fueled scrapers

Source: Photo taken by Bill Jacobus, http://en.wikipedia.org/wiki/File:Scraper.jpg

❖ Tractor/Loader/Backhoe

- ➢ Most often referred to as a "backhoe"
- ➢ Common and ubiquitous multipurpose equipment that includes the combined functions of a small loader and a small excavator in one unit.
- ➢ Source Classification Code is 2270002066 for diesel-fueled backhoes

Source: Photo taken by Radomil, http://en.wikipedia.org/wiki/File:Koparko_ladowarka.JPG

❖ **Skid Steer Loaders**
  ➢ Smaller (able to be 'skid' mounted to transport to a job site) loaders which may have alternative attachments than a bucket
  ➢ Source Classification Code is 2270002072 for diesel-fueled skid steer loaders

❖ Cranes
- ➢ Self-propelled
- ➢ Typically cable hoists
- ➢ Not to be confused with highway trucks with crane attachments runnin off the highway engine
- ➢ Source Classification Code is 2270002045 for diesel-fueled cranes

Source: http://en.wikipedia.org/wiki/File:Truck_crane.jpg

❖ **Excavators**
  ➢ Single-purpose wheeled or tracked excavators
  ➢ Consists of a boom, a bucket (or other attachment), and a cab on a rotating platform[12]
  ➢ Source Classification Code is 2270002036 for diesel-fueled excavators

Source: Photo taken by Alindon, 12 May 2009, http://en.wikipedia.org/wiki/File:LinkBelt290X2Excavator.jpg

---

[12] http://en.wikipedia.org/wiki/Excavator

❖ Bore/Drill Rigs
- ➤ Drills or boring rigs of all types that are skid-mounted, trailer-mounted, or self-propelled
- ➤ Not to be confused with highway trucks with drill attachments running off the highway engine
- ➤ Source Classification Code is 2270002033 for diesel-fueled bore/drill rigs

Source: http://en.wikipedia.org/wiki/File:RC_Drill_Rig_Western_Australia.jpg

❖ **Trenchers**
  ➤ Large and small trenchers typically use a rotating front-mounted rotating 'blade' to pull material from a trench and distribute it to the side
  ➤ Source Classification Code is 2270002030 for diesel-fueled trenchers

Source: Photo taken by Ky MacPherson, 25 June 2006, http://en.wikipedia.org/wiki/File:Trencher_2006-06-25.km.jpg

❖ Crawler Tractors/Dozers[13]

> ➤ A crawler tractor equipped with a blade used to push large quantities of material
> ➤ Typically equipped at the rear with a claw-like device known as a ripper to loosen densely-compacted materials
> ➤ Source Classification Code is 2270002069 for diesel-fueled dozers

---

[13] http://en.wikipedia.org/wiki/Bulldozer

❖ **Rough Terrain Forklifts (RTFs)**
   ➤ Often confused with typical, industrial forklifts or rubber tired loaders, but RTFs are specifically designed to operate off-road, having larger, knobby off-road tires and are specifically designed to handle palettes
   ➤ Include telescoping lift trucks called telescopic handlers
   ➤ Source Classification Code is 2270002057 for diesel-fueled RTFs

Source: Photo taken by Calibas, 14 November 2007, http://en.wikipedia.org/wiki/File:Telescopic_handler2.jpg

❖ **Graders**
  ➢ Also called road graders, motor graders, and maintainers
  ➢ Used to prepare a site, especially a road, for paving.
  ➢ A blade is mid-frame mounted
  ➢ Equipment has a long wheel-base
  ➢ Source Classification Code is 2270002048 for diesel-fueled graders

❖ **Off-Highway Trucks**
  ➢ Large, off-highway dump trucks not certified for highway use
  ➢ Source Classification Code is 2270002051 for diesel-fueled off-highway trucks

Source: Photo taken by Bidgee, http://en.wikipedia.org/wiki/File:Caterpillar_D350D.jpg

❖ Surfacing Equipment
  ➢ Various equipment use to supplement paving activity including paving material mixers, surface profilers (road reclaiming chippers), and seal coating equipment not used to distribute paving material as with paving equipment.
  ➢ Source Classification Code is 2270002024 for diesel-fueled surfacing equipment

❖ **Air Compressors**
  ➢ Trailer or skid mounted engine powered engine powered air compressors to generate high pressure air for pneumatic tools or other needs for pressurized air
  ➢ Source Classification Code is 2270006015 for diesel-fueled air compressors

❖ Generators
➢ Trailer or skid mounted self-contained engine/electric generator designed to supply electrical power at a job site
➢ Source Classification Code is 2270006005 for diesel-fueled generators.

Source: http://en.wikipedia.org/wiki/File:Big_boy.jpg

Check Digit Calculation[14]

First, find the numerical value associated with each letter in the VIN. (I, O and Q are not allowed.) Numerical digits use their own values.

A: 1	J: 1	
B: 2	K: 2	S: 2
C: 3	L: 3	T: 3
D: 4	M: 4	U: 4
E: 5	N: 5	V: 5
F: 6		W: 6
G: 7	P: 7	X: 7
H: 8		Y: 8
	R: 9	Z: 9

Second, look up the weight factor for each position in the VIN except the 9th (the position of the check digit).

1st: ×8 5th: ×4 10th: ×9 14th: ×5

2nd: ×7 6th: ×3 11th: ×8 15th: ×4

3rd: ×6 7th: ×2 12th: ×7 16th: ×3

4th: ×5 8th: ×10 13th: ×6 17th: ×2

Third, multiply the numbers and the numerical values of the letters by their assigned weight factor, and sum the resulting products. Divide the sum of the products by 11. The remainder is the check digit. If the remainder is 10, the check digit is the letter X. Valid check digits also run through the numbers zero to 9.

Example Check Digit Calculation

Consider the hypothetical VIN 1M8GDM9A_KP042788, where the underscore will be the check digit.

```
     VIN:  1  M  8  G  D  M  9  A  _  K  P  0  4  2  7  8  8
   Value:  1  4  8  7  4  4  9  1  0  2  7  0  4  2  7  8  8
  Weight:  8  7  6  5  4  3  2 10  0  9  8  7  6  5  4  3  2
Products:  8 28 48 35 16 12 18 10  0 18 56  0 24 10 28 24 16
```

The sum of all 16 products is 351. Dividing by 11 gives a remainder of 10, so the check digit is "X" and the complete VIN is 1M8GDM9AXKP042788.

---

[14] Example calculation taken from http://www.vinguard.org/vin.htm .

Appendix D
Average Fuel Consumption

Horsepower Range	Fuel Consumption (gallons/year)
3 < HP <= 6	154
6 < HP <= 11	240
11 < HP <= 16	395
16 < HP <= 25	603
25 < HP <= 40	950
40 < HP <= 50	1,290
50 < HP <= 75	1,762
75 < HP <= 100	2,471
100 < HP <= 175	3,626
175 < HP <= 300	6,616
300 < HP <= 600	12,037
600 < HP <= 750	19,939
750 < HP <= 1000	24,831
1000 < HP <= 1200	32,262
1200 < HP <= 2000	48,312
2000 < HP <= 3000	71,679

Appendix E
On-road Vehicle Identification Numbers

Decoding the Vehicle Identification Number (VIN)

Each position in a VIN has significance. The following is an example VIN.

1 F V A C W D C X 5 H U 8 6 2 6 6

The first three positions together are referred to as the World Manufacturer Identification (see Table D-1 for a list of common WMIs) and convey the following information.

Character 1 – Identifies the country in which the vehicle was manufactured
Character 2 – Identifies the manufacturer
Character 3 – Identifies the vehicle type or manufacturing division

Table E-1. Common WMIs for Heavy Duty Diesel Vehicle Manufacturers

WMI	Manufacturer
LVS	Ford Chang An
WF0	Ford Germany
1FA	Ford Motor Company
1FB	Ford Motor Company
1FC	Ford Motor Company
1FD	Ford Motor Company
1FM	Ford Motor Company
1FT	Ford Motor Company
8AF	Ford Motor Company Argentina
6F	Ford Motor Company Australia
9BF	Ford Motor Company Brazil
2FA	Ford Motor Company Canada
2FB	Ford Motor Company Canada
2FC	Ford Motor Company Canada
2FM	Ford Motor Company Canada
2FT	Ford Motor Company Canada
3FE	Ford Motor Company Mexico
VS6	Ford Spain
1FU	Freightliner
1FV	Freightliner
2FU	Freightliner
2FV	Freightliner
JF	Fuji Heavy Industries (Subaru)
2G	General Motors Canada
3G	General Motors Mexico
1G	General Motors USA
1GT	General Motors USA
6H	General Motors-Holden
JH	Honda
93H	Honda Brazil
2HG	Honda Canada
2HK	Honda Canada
3H	Honda Mexico

WMI	Manufacturer
SHS	Honda UK
1H	Honda USA
5F	Honda USA-Alabama
JA	Isuzu
LZE	Isuzu Guangzhou
1XK	Kenworth USA
1M1	Mack Truck USA
1M2	Mack Truck USA
1M3	Mack Truck USA
1M4	Mack Truck USA
A3	Mitsubishi
6MM	Mitsubishi Motors Australia
JN	Nissan
3N	Nissan Mexico
VSG	Nissan Spain
SJN	Nissan UK
1N	Nissan USA
5N1	Nissan USA
1XP	Peterbilt USA
4V1	Volvo
4V2	Volvo
4V3	Volvo
4V4	Volvo
4V5	Volvo
4V6	Volvo
4VL	Volvo
4VM	Volvo
4VZ	Volvo
YV3	Volvo Buse
YV1	Volvo Cars
YV2	Volvo Trucks
2WK	Western Star
2WL	Western Star
2WM	Western Star
2FZ	Sterling

Characters 4 through 8 – Positions four through eight are referred to as the vehicle descriptor section. They identify certain attributes of the vehicle such as body style, engine type, model, etc.

Character 9 – This is the "check digit" of the VIN for 1981 and newer model years. This number is used to verify the accuracy of any transcription of the vehicle VIN. The check digit value is determined by using a mathematical formula, as described in the "Check Digit Calculation" below.

Character 10 – Identifies the model year (See Table D-2)

Table E-2. Model Year Codes

Code	Year	Code	Year
A	1980	Y	2000
B	1981	1	2001
C	1982	2	2002
D	1983	3	2003
E	1984	4	2004
F	1985	5	2005
G	1986	6	2006
H	1987	7	2007
J	1988	8	2008
K	1989	9	2009
L	1990	A	2010
M	1991	B	2011
N	1992	C	2012
P	1993	D	2013
R	1994	E	2014
S	1995	F	2015
T	1996	G	2016
V	1997	H	2017
W	1998	J	2018
X	1999	K	2019

*Note: The letters U and Z and the digit 0 are not used for the year code.

Character 11 – Identifies the vehicle assembly plant.

Characters 12 through 17 – Positions twelve through seventeen are referred to as the vehicle identification section. They indentify the specific vehicle. The last four digits will always be numeric and identifies the sequence of the vehicle for production as it was rolled off of the assembly line.

Appendix F
Standardized Engine Family and Evaporative Family Names for 1997 and Earlier Model Years

Character Meaning

First Character — Model Year (See Model Year Subcodes in Table E-1)

Characters 2 & 3 — Letter Code for Manufacturer (See Manufacturer Subcodes in Table E-2)

Characters 4, 5, & 6 — Displacement in Liters or Cubic Inches (If one of these digits is represented by a decimal place, the displacement is presented in liters. Otherwise, the displacement is in cubic inches and can be converted to liters by multiplying by 0.0164.)

Character 7 — Vehicle Class (See the Vehicle Classes in Tables E-3 through E-6)

Character 8 — Fuel System and number of valves (See Table E-7)

Character 9 — Combustion Cycle and Fuel (See Table E-8)

Character 10 — Standards (See Table E-9)

Character 11 — Catalyst, FFS (See Table E-10)

Character 12 — Emission Control Devices (or ICI Production Year) (See Table E-11 through E-14)

Model Year

The following table provides the model year codes that are seen in the engine family name. Use the code in the first character of the engine family name to find the correct model year for the engine.

Table F-1. Subcodes for Model Year

CODE	YEAR	CODE	YEAR	CODE	YEAR
A	1980	M	1991	2	2002
B	1981	N	1992	3	2003
C	1982	P	1993	4	2004
D	1983	R	1994	5	2005
E	1984	S	1995	6	2006
F	1985	T	1996	7	2007
G	1986	V	1997	8	2008
H	1987	W	1998	9	2009
J	1988	X	1999	A	2010
K	1989	Y	2000	B	2011
L	1990	1	2001	C	2012

Table F-2. Subcodes for Manufacturers

MFR Code	Product	Manufacturer or Laboratory	Manufacturer/lab Subcodes	
			<1994	<1997
10	LD	CHRYSLER (AMC)	AM	same
20	LD	CHRYSLER	CR	same
	HD	CHRYSLER	CC2	CR
30	LD	FORD	FM	same
	HD	FORD	FM	same
40		GENERAL MOTORS	GC	GC,GM3
	LD	CPC (Chevrolet, Pontiac)	IG	1G.GM
	LD	BUICK-OLDSMOBILE-CADILLAC	2G	2G,GM
	HD	TRUCK & BUS	30	3G,GM
	LD	SATURN	40	4G,GM
	HD	GENERAL MOTORS	GM	GM
52	LD,UE,HD,SN,MC,LN,IL,GL,ME			
		TASMANIA MOTOR WORKS4	TW	same
55	HD	DETROIT DIESEL	DD	same
60	LD	AC CARS LIMITED	ZZ	same
67	LD	AMERICAN LIMOUSINE MFR. INC	Z6	same
68	LD	AMERICAN MUSCEL LTD	A4	same
69	HD	AMERICAN TECHNOLOGY GROUP	A9	same
70	LD	ASTON MARTIN	AS	same
90	LD	FIAT AUTO S.P.A.	AR	same
95	HD	AM GENERAL	AZ	same
98	LD	AURORA CARS	AA	same
101	LD	AUTOKRAFT LIMITED	AK	same
103	LD	ASC INC.	A3	same
106	LD	ALLCO EURO MOTORS	A6	same
108	LD	ROVER GROUP LTD. (AR)	AW	same
112	HD	BLUE BIRD BODY	BB	same
118	MC	BAJAJ AUTO LIMITED	BO	same
119	MC	BUELL MOTORCYCLE	BL	Same
120	LD	BMW	BM	same
	MC	BMW AG	BM	same
123	MC	BIMOTA S.P.A.	Z8	same
126	LD	BONAIR USA	B3	same
133	LN	BAKER EQUIPMENT ENGINEERING CO.	X3	same
134	LD	BUGATTI AUTOMOBILI SPA	BA	same
141	LD	CHAMPAGNE IMPORTS INC.	Z5	same
143	LD	CALLAWAY	C6	same

MFR Code	Product	Manufacturer or Laboratory	Manufacturer/lab Subcodes	
			<1994	<1997
144	MC	CAGIVA NORTH AMERICA	CG	same
146	LD	CHICAGO ARMOR&LIMOUSINE MFR CORP	Z7	same
147	LD	CCE, INC	C7	same
150	LD	CITROEN	CT5	same
156	MC	CLASSIC MOTORCYCLES LIMITED	CM	same
157	MC	CLIFFORD GUN TRADERS & SUPPLIES	CL	same
162	LD	CONSULIER INDUSTRIES INC.	C3	same
163	LD	COLLINS PROFESSIONAL CARS. INC.	Y4	same
168	MC	CUSHMAN	CU	CH
169	LD	CX AUTOMOTIVE	CX	same
175	LD	DACIA (ARO)	DA	same
178	LD	DAEWOO	DW	same
180	HD	DAF	DT6	DF
185	LD	DABRYAN COACH BUILDERS INC.	Y2	same
190	LD	DAIHATSU MOTOR COMPANY LTD.	DH	same
196	LD	MITSUBISHI MOTOR MANUF OF AMERICA	DS	same
197	LD	DUTCHER MOTORS INC	DT'	same
200	LD	MERCEDES BENZ	MB	same
	HD	MERCEDES-BENZ AKTIENGELLSCHAFT	MB	same
201	LD	EMPIRE COACH	E6	same
204	LD,HD	US ELECTRICAR	EL	same
206	LD	DNIEPER U.S.A.	DP	same
207	LD	EXECUTIVE COACH BUILDERS	Y3	same
208	LD	ECS/ROUSH	E5	same
212	LD	EUROPEAN AUTO WERKS, INC.	E2	same
220	LD	FERRARI	FE	same
222	LD	EVANS AUTOMOBILES	El	same
227	LD	FEDERAL COACH	F2	same
230	LD	FIAT	FT6	same
241	HD	FREIGHTLINER	FR	same
242	LD	GREEN WHEELS ELECTRIC	G4	same
243	MC	ALEX GREENSPAN T/A FIN	GA	same
244	LD	GREENWOOD AUTOMOTIVE PERFORMANCE	GW	same
246	LD	GRUMMAN ALLIED INDUSTRIES	GR	same
250	HD	HINO MOTORS	HM	same
251	LD	G & K AUTOMOTIVE CONVERSION INC	G1	same

MFR Code	Product	Manufacturer or Laboratory	Manufacturer/lab Subcodes	
			<1994	<1997
253	LD	VECTOR AEROMOTIVE CORPORATION	G2	same
254	LD	GOLDACRE LTD.	G3	same
255	MC	HARLEY DAVIDSON	HD	same
258	SN	HATZ GMBH & CO KG	HZ	same
260	LD	HONDA	HN	Same
	MC	HONDA	HN	same
	UE	HONDA	HN	same
265	LD	HYUNDAI	HY	same
266	LD	ICI-INTERNATIONAL	X1	same
271	LD	IMPCO	Z9	same
272	LD	IMPORT TRADE SERVICES	TI	same
285	LD	ISIS IMPORTS LTD	Z3	same
290	LD	ISUZU	SZ	same
	HD	ISUZU MOTORS	SZ	same
305	LD	JAGUAR CARS INC JR (WAS JC) JC		
308	LD	JBA MOTORCARS INC	J1	same
314	LD	J.K. MOTORS	J3	same
329	LD	KINGS ENVIRONMENTAL HYDROGEN SYS	K4	same
331	MC	KAVULICH INTERNATIONAL	MN	M5
332	LD	KRYSTAL COACH INC.	KK	same
333	MC	KTM MOTOR	KT	same
335	MC	KAWASAKI	KA	same
338	LD	KIA MOTORS CORPORATION	KM	same
339	LD	KSK DISTRIBUTING	K2	same
344	LD	LIMOUSINE WERKS	L6	same
347	LD	LIPHARDT & ASSOCIATES INC	LP	same
350	LD	LOTUS	LT	same
352	LD	LAREDO COACHWORKS, INC	L7	same
355	HD	STEELBRO MANUFACTURING, LTD	SB	same
357	SN	MAKITA USA INC	M6	same
358	MC	MATCHLESS MOTOR CYCLES	MA	M2
360	LD	MASERATI	MA	same
366	MC	MILLER SPECIALTIES	MS	same
369	MC	MOTO AMERICA	MG	same
371	MC	MUZ, MOTORRAD UND ZWEIRADWERK	MZ	same
373	LD	NORTH AMERICAL MVS	N3	same
374	MC	NATIVE AMERICAN MOTORCYCLE CO.	N6	same

MFR Code	Product	Manufacturer or Laboratory	Manufacturer/lab Subcodes	
			<1994	<1997
376	LD	NEOAX	NX	same
378	MC	NEVAL MOTORCYCLES	NL	NY
380	LD	NISSAN	NS	same
381	HA	NISSAN DIESEL MOTOR CO.	ND	same
394	MC	OMC LINCOLN	MC	same
404	LD	PRODUCTION AUTOMOTIVE SYSTEMS	P5	same
407	LD	PANOZ AUTO-DEVELOPMENT CORP	P3	same
410	LD	PEUGEOT	PE	same
416	LD	PIERRE ENTERPRISES SOUTHEAST, INC	P5	same
420	LD	PORSCHE	PR	same
426	LA	PYRAMID COACHBUILDERS	P4	same
430	LA	RENAULT	RE	same
431	LD	PAS INC.	P2	same
432	LA	RENNTECH INC.	R2	same
433	HD	RENAULT VEHICLES INDUSTRIELS	R3	same
439	LD	RAYTON-FISSORE NORTH AMERICA	R1	same
440	LD	ROLLS-ROYCE MOTORCARS LTD.	RR	same
453	MC	ROSCETTI	RC	same
454	LD	RUF AUTOMOBILE GMBH	RA	same
457	LA	ROYALE LIMOUSINE MANUFACTURERS	RL	same
460	LD	ROVER GROUP LTD.	LR	same
470	LD	SAAB	SA	same
	HD	SAAB SCANIA	SS	SA
471	LD	SAAC CAR COMPANY INC.	S6	same
472	LA	SALEEN AUTOSPORT	S3	same
473	LD	SALEEN PERFORMANCE PARTS, INC.	S8	same
475	LA	SEGUELS SERVICE INC	S2	same
481	LD	SHELBY AUTOMOBILES INC	SY	same
487	LD	SLP ENGINEERING	S5	same
490	LD	MITSUBISHI	MT	same
	HD	MITSUBISHI	MM	MT
491	LA	MITSUBISHI MOTOR SALES AMERICA	M3	same
492	LD	MITSUBISHI MOTORS AUSTRALIA LTD	ML	same
515	LD	SUPERIOR OF OHIO INC	V1	same
520	LD	EXCALIBUR AUTOMOBILE	EX	same
526	LD	TDM TECHNOLOGIES, INC.	T4	same

MFR Code	Product	Manufacturer or Laboratory	Manufacturer/lab Subcodes	
			<1994	<1997
527	LA	THOMAS PUGH AND LINDA MCKNIGHT	T3	same
529	HD	TRANSI-CORP	T5	same
530	MC	TRIUMPH DESIGNS LTD	TD	same
534	LA	SPORTS CAR AMERICA PUMA DIVISION	Z4	same
540	LA	SUZUKI MOTOR CORPORATION	SK	same
560	LD	MAZDA MOTOR CORP.	TK	same
570	LD	TOYOTA	TY	same
576	LD	NEW UNITED MOTOR MFG INC	NT	same
579	LD	UTILIMASTER CORP. OF AMERICA	Z1	same
581	MC	URALMOTO JSC	YP	same
582	LD	UNITED STATES COACHWORKS	Y6	same
583	LD	US TRADE CORP.	Z2	same
590	LD	VOLKSWAGEN	VW	same
600	LD	VOLVO	VV	same
603	LD	WALLACE ENVIR.TESTING LAGS. INC	WA	same
605	HD	VOLVO WHITE TRUCK DIVISION	VT	same
608	LD	WISCONSIN LIFT TRUCK CORP.	WL	same
611	MC	WESTWARD INDUSTRIES	WW	same
614	LD	YUGO AMERICA, INC.	YA	same
615	MC	YAMAHA	YA	YM
640	LD	AUDI	AD	same
645	LD	AMPHI-RANGER OF AMERICA	Y1	same
660	LD	FUJI HEAVY IND	FJ	same
661	SN	FUJI ROBIN INDUSTRIES LTD.	FN	same
691	LD	LAMBORGHINI	NL	same
720	HD	WINNEBAGO INDUSTRIALS	WB	same
728	HD	ASQUITH MOTOR CARRIAGE CO. LTD	A7	same
730	HD	CATERPILLER	CT	CP
735	HD	CLARION MOTORS	CA	same
	MC	CLARION MOTORS	CA	same
740	HD	CUMMINS	CE	same
743	HD	DEERE & COMPANY	JD	same
745	HD	KLOCKNER-HUMBOLT-DEUTZ AG	DZ	same
747	HD	FLEETWOOD ENTERPRISES	FW	same
748	HD	GILLIG	GL	same
750	HD	HERCULES ENGINES	HE	same
755	HD	IVECO B.V.	VE	same

MFR Code	Product	Manufacturer or Laboratory	Manufacturer/lab Subcodes	
			<1994	<1997
760	HD	MACK TRUCKS	MT	MK
762	HD	MAN NUTZPAHRZEUGE	MN	same
765	HD	NAVISTAR INTERNATIONAL TRANS.	NV	same
767	HD	OSHKOSH TRUCK	FT	S7
770	HD	PERKINS ENGINE COMPANY	PE	PK
775	HD	ROADMASTER	RM	same
777	LD,UE,HD,SN,MC,LN,IL,GL,ME			
		JURASSIC PASSENGER CARS27	JP	same
793	HD	TRANSPORTATION MANUFACTURING COR	T6	same
795	HD	VIRONEX	VX	same
802	UE	ANDREAS STIHL	A8	same
805	UE	BRIGGS & STRATTON	BS	same
815	LN	DAE HUNG	DE	same
825	UE	KIORITZ	EH	same
828	UE	GENERAC CORP	GN	same
835	UE	HOMELITE TEXTRON	H2	same
838	UE	HUSQVARNA AB	HV	same
840	UE	INERTIA DYNAMICS CORP.	N4	same
845	UE	KOHLER COMPANY	KH	same
847	UE	KOMATSU ZENOAH AMERICA	KZ	same
848	LN	KOMATSU LTD.	KL	same
849	UE	KUBOTA	KB	same
850	UE	LAWN-BOY	L4	same
852	UE	LISTER PETTER, INC.	L5	same
854	SN	MARUYAMA U.S. INC	M4	same
855	UE	MCCULLOCH CORP.	MH	same
860	UE	NELSON	NE	same
865	UE	ONAN CORP	N5	same
867	SN	SOLO INC	S9	same
868	UE	POULAN/WEED EATER	PW	same
869	UE	SHINDAIWA INC	SW	same
870	UE	TECUMSEH PRODUCTS	TP	same
871	SN	TANAKA KOGYO CO LTD	T7	same
872	UE	TELEDYNE TOTAL POWER	T2	same
885	UE	YANMAR DIESEL ENGINE USA	YD	same
890	UE	WACKER CORP.	W1	same
893	SN	WIS-CON TOTAL POWER CORP	WP	same
901	IL	AUTOMOTIVE TESTING LABS, INC.	1	same
902	IL	ECS LABORATORIES INC.	2	same

MFR Code	Product	Manufacturer or Laboratory	Manufacturer/lab Subcodes	
			<1994	<1997
903	IL	ENVIRONMENTAL TESTING CORP.	3	same
904	IL	LUCAS ENGINE MANAGEMENT SYSTEMS	4	same
905	IL	ENVIRONMENTAL RESEARCH & DEV. CO	5	same
906	IL	NORTHERN CAL.EMISSIONS LAB.	6	same
907	IL	TESTING SERVICES INC.	7	same
908	IL	COMPLIANCE & RESEARCH SERVICES	8	same
909	IL	AUTOMATED CUSTOM SYSTEMS, INC.	9	same
910	IL	CALIFORNIA ENVIRONMANTAL ENG.	10	same
911	IL	EAGLE PITCHER AUTOMOTIE GROUP	11	same
912	IL	TICKFORD LIMITED	K3	same
920	GL	COUNTRY OF SWEDEN	SG	same
980	GL	CALIFORNIA AIR RESOURCES BOARD	80	same
991	GL	EPA CD	91	same
992	GL	EPA EOD	92	same
993	GL	EPA MOD	93	same
994	GL	EPA FOSD	94	same
995	GL	EPA ECTD (obsolete)	95	same
996	GL	EPA RDSD	96	same
997	GL	EPA EPSD	97	same

Table F-3. Light-Duty Vehicle Classification Codes

CODE	LDV or CARB's PC	GVW	TIER1	TIER 0
1	3,750	6,000	LDT1	LDT-A-NOx 1.2
2	>3,750	6,000	LDT2	LDT-B -NOx 1.7
3	3,750	>6,000	LDT3	LDT-A-NOx 1.2
4	>3,750	>6,000	LDT3	LDT-B -NOx 1.7
5	3,750	>6,000	LDT4	LDT-A-NOx 1.2
6	>3,750	>6,000	LDT4	LDT-B-NOx 1.7

Table F-4. Medium-Duty Vehicle Classification Codes (CARB)

CODE	DESIGNATION	GVWR
H	MDT-1	>6,000
J	MDT-2	>6,000
K	MDT-3	>6,000
L	MDT-4	>6,000
M	MDT-5	>6,000

Table F-5. Heavy-Duty Vehicle Classification Codes

CODE	STANDARD	DESCRIPTION
A	LIGHT-DUTY	OPTION for <10,000 GVW
B	<14K GVW	Typically GVW <19.5K, HP 70-170
C	>14K GVW	Typically GVW <19.5K, HP 70-170
D	>14K GVW	Typically GVW 19.5K -33K, HP 170-250
E	>14K GVW	Typically GVW >33K, HP >250
F	HHDE Bus	
G	Vehicle Evap Compliance	

Table F-6. Miscellaneous Classification Codes

U CARB'S UTILITY ENGINE & LAWN/GARDEN

Table F-7. Fuel Metering and Valves per Cylinder

CODE	FUEL SYSTEM	VALVES PER CYLINDER
O	Multiple Carburetor	2 Valves/Cylinders
1	1 Barrel Carburetor (BBL)	2 Valves/ Cylinders
2	2 Barrel Carburetor (BBL)	2 Valves/ Cylinders
3	3 Barrel Carburetor (BBL)	2 Valves/Cylinders
4	4 Barrel Carburetor (BBL)	2 Valves/ Cylinders
5	Throttle Body Injection (TBI)	2 Valves/ Cylinders
6	Mechanical Multi-Point Injection (MPI)	2 Valves/Cylinders
7	Electric Multi-Point Injection (MPI)-simultaneous	2 Valves/Cylinders
8	Electric Multi-Point Injection (MPI)-sequential	2 Valves/Cylinders
9	Central Port Injection	2 Valves/Cylinders
A	Multiple Carburetor	3 or more Valves/Cylinders
B	1 Barrel Carburetor (BBL)	3 or more Valves/Cylinders
C	2 Barrel Carburetor (BBL)	3 or more Valves/Cylinders
D	3 Barrel Carburetor (BBL)	3 or more Valves/Cylinders
E	4 Barrel Carburetor (BBL)	3 or more Valves/Cylinders
F	Throttle Body Injection (TBI)	3 or more Valves/Cylinders
G	Mechanical Multi-Point Injection (MPI)	3 or more Valves/Cylinders
H	Electric Multi-Point Injection (MPI)-simultaneous	3 or more Valves/Cylinders
J	Electric Multi-Point Injection (MPI)-sequential	3 or more Valves/Cylinders
K	Central Port Injection	3 or more Valves/Cylinders
Z	Other	

Table F-8. Combustion Cycle and Fuel

CODE	CYCLE	FUEL	ENGINE TYPE
G	Otto Cycle (SI)	Gasoline	Piston
M	Otto Cycle (SI)	Methonal	Piston
E	Otto Cycle (SI)	Ethanol	Piston
F	Otto Cycle (SI)	Flexible Methanol-Gasoline	Piston
N	Otto Cycle (SI)	Other Flexible	Piston
C	Otto Cycle (SI)	CNG	Piston
L	Otto Cycle (SI)	LPG	Piston
R	Otto Cycle (SI)	Gasoline	Rotary
X	Otto Cycle (SI)	Other Fuels	Rotary
D	Diesel Cycle (CI)	Diesel Fuel	
A	Diesel Cycle (CI)	Methonal	
B	Diesel Cycle (CI)	Ethanol	
H	Diesel Cycle (CI)	Flexible Methanol-Gasoline	
J	Diesel Cycle (CI)	Other Flexible	
K	Diesel Cycle (CI)	CNG	
P	Diesel Cycle (CI)	LPG	
2	Two Stroke Cycle	Gasoline	
3	Two Stroke Cycle	Methonal/Ethanol	
4	Two Stroke Cycle	Diesel	
5	Two Stroke Cycle	CNG	
6	Two Stroke Cycle	LPG	
7	Two Stroke Cycle	Flexible	
T	Turbine	Gasoline	
Q	Turbine	Diesel	
S	Turbine	Methonal/Ethanol	
U	Turbine	CNG	
V	Turbine	LPG	
W	Turbine	Flexible	
Y	Hybred Electric		
Z	Electric		

Table F-9. Family Standards Tier Designation Codes

CODE	SALES CLASS	HC, CO, & NOx	PM	EVAP	COLD CO	IN USE
A	49 OR 50 STATE	TIER 0	ANY	TIER 0	N	TIER 0
B	49 OR 50 STATE	TIER 0	ANY	TIER 0	Y	TIER 0
C	49 OR 50 STATE	TIER 1	TIER 0	TIER 0	N	TIER 1l
D	49 OR 50 STATE	TIER 1	TIER 0	TIER 0	Y	TIER 1l
E	49 OR 50 STATE	TIER 1	TIER 1	TIER 0	N	TIER 1l
F	49 OR 50 STATE	TIER 1	TIER 1	TIER 0	Y	TIER 1l
G	49 OR 50 STATE	TIER 1	TIER 0	TIER 0	N	TIER 1F
H	49 OR 50 STATE	TIER 1	TIER 0	TIER 0	Y	TIER 1F
J	49 OR 50 STATE	TIER 1	TIER 1	TIER 0	N	TIER 1F
K	49 OR 50 STATE	TIER 1	TIER 1	TIER 0	Y	TIER 1F
L	CLEAN FUELS FLEET					
M	NCP					
N	AVE OR BANK/TRADE					
P-Z	(RESERVED)					
O	(RESERVED)					
CALIFORNIA ONLY FAMILIES						
	1 CARB TIER1					
	2 CARB TLEV					
	3 CARB LEV					
	4 CARB ULEV					
	5 CARB ZEV (ELECRIC)					
Note: Exact standards can usually be determined knowing the class of vehicle and the year of certification. However, for some years there are more than one standard effective and there are phase-in percentages required. The "standard" in the above table identifies which standard applies.						
Tier 0						
LDV, LDT:	As defined in regulations					
HDE:	Standards through 1997					
MC:	Current Standards					
Tier 1						
LDV, LDT:	As defined in regulations					
HDE:	1998 standards and later					
MC:	Not applicable					

Table F-10. Catalyst / OBD Codes

CODE	CATALYST TYPE	MATERIAL	FEDERAL OBD	CARB OBD
A	Ox Cat Only	Any	N	
B	Ox Cat Only	Any	Y	II
C	Reduction Cat	Any	N	
D	Reduction Cat	Any	Y	II
E	3-Way Cat	Ceramic Monolyth	N	
F	3-Way Cat	Ceramic Monolyth	Y	II
G	3-Way Cat	Pellets	N	
H	3-Way Cat	Pellets	Y	II
J	3-Way Cat	Metal	N	
K	3-Way Cat	Metal	Y	II
L	3-Way Cat	Other or Mixed	N	
M	3-Way Cat	Other or Mixed	Y	II
N	3-Way+Ox Cat	Ceramic Monolyth	N	
P	3-Way+Ox Cat	Ceramic Monolyth	Y	II
Q	3-Way+Ox Cat	Pellets	N	
R	3-Way+Ox Cat	Pellets	Y	II
S	3-Way+Ox Cat	Metal	N	
T	3-Way+Ox Cat	Metal	Y	II
U	3-Way+Ox Cat	Other or Mixed	N	
V	3-Way+Ox Cat	Other or Mixed	Y	II
W	Heated Cat	Any	N	
X	Heated Cat	Any	Y	II
Y	No Cat	Any	N	
Z	No Cat	Any	Y	II

Table F-11. Trap Type Codes

CODE	TRAP TYPE	FEDERAL OBD	CARB OBD
1	Trap -Active Regeneration	N	I
2	Trap -Active Regeneration	Y	II
3	Trap-Continous Regeneration	N	I
4	Trap-Continous Regeneration	Y	II
5	Trap-Continous Regeneration + Fuel Add.	N	I
6	Trap-Continous Regeneration + Fuel Add.	Y	II

Table F-12. OBD Codes

CODE	DESCRIPTION	FEDERAL OBD	CARB OBD
9	Other	N	I
O	Other	Y	II

Table F-13. Emission Control System Codes* – ICI Prod Year

A,B,C	EGR [and other]
D,E,F	EGR + Air [and other]
G,H,J	EGR + T/C or S/C [and other]
K,L,M	EGR + Air + T/C or S/C [and other]
N,P,Q	Air [and other]
R,S,T	Air + T/C or S/C [and other]
U,V,W	T/C [and other]
X,Y,Z	S/C [and other]
6,7	Other Only
8,9	NONE

* First code listed is preferred code, other codes may be
selected if necessary to separate engine families that would
otherwise be named the same.

Table F-14. ICI Codes

Codes	Production Year
5	Production year is 5 years earlier than certified model year
4	Production year is 4 years earlier than certified model year
3	Production year is 3 years earlier than certified model year
2	Production year is 2 years earlier than certified model year
1	Production year is 1 year earlier than certified model year
0	Production year is same year as certified model year

Appendix G
Standardized Engine and Evaporative Family Names
for 1998 and Later Model Years

Table G-1. Family Type code for All families
(Exhaust or Evaporative) or Test Groups

A	California only medium duty engine family or test groups
B	**Test Group consisting of:**
	Both light-duty vehicle (LDV) and light-duty truck (LDT); or
	Both LDV and medium-duty passenger vehicle (MDPV)
C	Motorcycle engine family
E	Evaporative family
H	Heavy-duty engine family or test groups
K	Complete heavy duty vehicle (tested on chassis dynamometer)
L	Large nonroad family
M	Marine engine family
N	Nonstandard family type
R	Evaporative/Refueling family
S	Small nonroad family
T	Light-duty truck engine family or test group
V	Light-duty vehicle engine family or test group
X	**Off-highway motorcycles (OHMC) and all-terrain vehicles (ATV)**

Table G-2. Letter Codes for Model Year

Code	Year	Code	Year	Code	Year
A	1980	L	1990	Y	2000
B	1981	M	1991	1	2001
C	1982	N	1992	2	2002
D	1983	P	1993	3	2003
E	1984	R	1994	4	2004
F	1985	S	1995	5	2005
G	1986	T	1996	6	2006
H	1987	V	1997	7	2007
J	1988	W	1998	8	2008
K	1989	X	1999	9	2009

Table G-3. Engine/Evaporative Family Manufacturer Subcodes and Vehicle Information Manufacturer Codes

LD	-	Light-duty vehicles	UE	-	Utility engines
HD	-	Heavy-duty vehicles/engines	SN	-	Small nonroad engines
MC	-	Motorcycles	LN	-	Large nonroad engines
IL	-	Independent testing lab.	GL	-	Government laboratory
ME	-	Marine engines			

MFR Code	Product	Manufacturer or laboratory	Manufacturer/lab Subcodes		
			<1994	<1997	≥1998
10	LD	CHRYSLER (AMC)	AM	same	AMX
20	LD	CHRYSLER	CR	same	CRX
	HD	CHRYSLER	CC2	CR	CRX
30	LD	FORD	FM	same	FMX
	HD	FORD	FM	same	FMX
40		GENERAL MOTORS	GC	GC,GM3	GCX
	LD	CPC (Chevrolet, Pontiac)	IG	1G.GM	GMX
	LD	BUICK-OLDSMOBILE-CADILLAC	2G	2G,GM	GMX
	HD	TRUCK & BUS	30	3G,GM	GMX
	LD	SATURN	40	4G,GM	GMX
	HD	GENERAL MOTORS	GM	GM	GMX
52	LD,UE,HD,SN,MC,LN,IL,GL,ME				
		TASMANIA MOTOR WORKS4	TW	same	TWX
55	HD	DETROIT DIESEL	DD	same	DDX
60	LD	AC CARS LIMITED	ZZ	same	zzx
67	LD	AMERICAN LIMOUSINE MFR. INC	Z6	same	Z6X
68	LD	AMERICAN MUSCEL LTD	A4	same	A4X
69	HD	AMERICAN TECHNOLOGY GROUP	A9	same	A9X
70	LD	ASTON MARTIN	AS	same	ASX
90	LD	FIAT AUTO S.P.A.	AR	same	ARX
95	HD	AM GENERAL	AZ	same	AZX
98	LD	AURORA CARS	AA	same	AAX
101	LD	AUTOKRAFT LIMITED	AK	same	AKx
103	LD	ASC INC.	A3	same	A3X
106	LD	ALLCO EURO MOTORS	A6	same	A6X
108	LD	ROVER GROUP LTD. (AR)	AW	same	AWX
112	HD	BLUE BIRD BODY	BB	same	BBX
118	MC	BAJAJ AUTO LIMITED	BO	same	B,71
119	MC	BUELL MOTORCYCLE	BL	Same	BLX
120	LD	BMW	BM	same	BMX
	MC	BMW AG	BM	same	BMX
123	MC	BIMOTA S.P.A.	Z8	same	Z8X

MFR Code	Product	Manufacturer or laboratory	Manufacturer/lab Subcodes		
			<1994	<1997	≥1998
126	LD	BONAIR USA	B3	same	B3X
133	LN	BAKER EQUIPMENT ENGINEERING CO.	X3	same	X3X
134	LD	BUGATTI AUTOMOBILI SPA	BA	same	BAX
141	LD	CHAMPAGNE IMPORTS INC.	Z5	same	Z5X
143	LD	CALLAWAY	C6	same	C6X
144	MC	CAGIVA NORTH AMERICA	CG	same	CGX
146	LD	CHICAGO ARMOR&LIMOUSINE MFR CORP	Z7	same	Z7X
147	LD	CCE, INC	C7	same	C7X
150	LD	CITROEN	CT5	same	CTX
156	MC	CLASSIC MOTORCYCLES LIMITED	CM	same	CMX
157	MC	CLIFFORD GUN TRADERS & SUPPLIES	CL	same	CLX
162	LD	CONSULIER INDUSTRIES INC.	C3	same	C3X
163	LD	COLLINS PROFESSIONAL CARS. INC.	Y4	same	Y4X
168	MC	CUSHMAN	CU	CH	CUX
169	LD	CX AUTOMOTIVE	CX	same	CXX
175	LD	DACIA (ARO)	DA	same	DAX
178	LD	DAEWOO	DW	same	DWX
180	HD	DAF	DT6	DF	DTX
185	LD	DABRYAN COACH BUILDERS INC.	Y2	same	Y2X
190	LD	DAIHATSU MOTOR COMPANY LTD.	DH	same	DHX
196	LD	MITSUBISHI MOTOR MANUF OF AMERICA	DS	same	DSX
197	LD	DUTCHER MOTORS INC	DT'	same	DTX
200	LD	MERCEDES BENZ	MB	same	MBX
	HD	MERCEDES-BENZ AKTIENGELLSCHAFT	MB	same	MBX
201	LD	EMPIRE COACH	E6	same	E6X
204	LD,HD	US ELECTRICAR	EL	same	ELX
206	LD	DNIEPER U.S.A.	DP	same	DPX
207	LD	EXECUTIVE COACH BUILDERS	Y3	same	Y3X
208	LD	ECS/ROUSH	E5	same	ESX
212	LD	EUROPEAN AUTO WERKS, INC.	E2	same	E2X
220	LD	FERRARI	FE	same	FEX
222	LD	EVANS AUTOMOBILES	El	same	E1X
227	LD	FEDERAL COACH	F2	same	F2X
230	LD	FIAT	FT6	same	FTX
241	HD	FREIGHTLINER	FR	same	FRX
242	LD	GREEN WHEELS ELECTRIC	G4	same	G4X
243	MC	ALEX GREENSPAN T/A FIN	GA	same	GAX
244	LD	GREENWOOD AUTOMOTIVE PERFORMANCE	GW	same	GWX
246	LD	GRUMMAN ALLIED INDUSTRIES	GR	same	GRX
250	HD	HINO MOTORS	HM	same	HMX
251	LD	G & K AUTOMOTIVE CONVERSION INC	G1	same	GlX

MFR Code	Product	Manufacturer or laboratory	Manufacturer/lab Subcodes		
			<1994	<1997	≥1998
253	LD	VECTOR AEROMOTIVE CORPORATION	G2	same	G2X
254	LD	GOLDACRE LTD.	G3	same	G3X
255	MC	HARLEY DAVIDSON	HD	same	HDX
258	SN	HATZ GMBH & CO KG	HZ	same	HZX
260	LD	HONDA	HN	Same	HNX
	MC	HONDA	HN	same	HNX
	UE	HONDA	HN	same	HNX
265	LD	HYUNDAI	HY	same	HYX
266	LD	ICI-INTERNATIONAL	X1	same	X1X
271	LD	IMPCO	Z9	same	Z9X
272	LD	IMPORT TRADE SERVICES	TI	same	T1X
285	LD	ISIS IMPORTS LTD	Z3	same	Z3X
290	LD	ISUZU	SZ	same	SZX
	HD	ISUZU MOTORS	SZ	same	SZX
305	LD	JAGUAR CARS INC JR (WAS JC) JC			JCX
308	LD	JBA MOTORCARS INC	J1	same	J1X
314	LD	J.K. MOTORS	J3	same	J3X
329	LD	KINGS ENVIRONMENTAL HYDROGEN SYS	K4	same	K4X
331	MC	KAVULICH INTERNATIONAL	MN	M5	MNX
332	LD	KRYSTAL COACH INC.	KK	same	KKX
333	MC	KTM MOTOR	KT	same	KTX
335	MC	KAWASAKI	KA	same	KAX
338	LD	KIA MOTORS CORPORATION	KM	same	KMX
339	LD	KSK DISTRIBUTING	K2	same	K2X
344	LD	LIMOUSINE WERKS	L6	same	L6X
347	LD	LIPHARDT & ASSOCIATES INC	LP	same	LPX
350	LD	LOTUS	LT	same	LTX
352	LD	LAREDO COACHWORKS, INC	L7	same	L7X
355	HD	STEELBRO MANUFACTURING, LTD	SB	same	SBX
357	SN	MAKITA USA INC	M6	same	M6X
358	MC	MATCHLESS MOTOR CYCLES	MA	M2	M2X
360	LD	MASERATI	MA	same	MAX
366	MC	MILLER SPECIALTIES	MS	same	MSX
369	MC	MOTO AMERICA	MG	same	MGX
371	MC	MUZ, MOTORRAD UND ZWEIRADWERK	MZ	same	MZX
373	LD	NORTH AMERICAL MVS	N3	same	N3X
374	MC	NATIVE AMERICAN MOTORCYCLE CO.	N6	same	N6X
376	LD	NEOAX	NX	same	NXX
378	MC	NEVAL MOTORCYCLES	NL	NY	NYX
380	LD	NISSAN	NS	same	NSX
381	HA	NISSAN DIESEL MOTOR CO.	ND	same	NDX
394	MC	OMC LINCOLN	MC	same	MCX

MFR Code	Product	Manufacturer or laboratory	Manufacturer/lab Subcodes		
			<1994	<1997	≥1998
404	LD	PRODUCTION AUTOMOTIVE SYSTEMS	P5	same	P5X
407	LD	PANOZ AUTO-DEVELOPMENT CORP	P3	same	P3X
410	LD	PEUGEOT	PE	same	PEX
416	LD	PIERRE ENTERPRISES SOUTHEAST, INC	P5	same	P5X
420	LD	PORSCHE	PR	same	PRX
426	LA	PYRAMID COACHBUILDERS	P4	same	P4X
430	LA	RENAULT	RE	same	REX
431	LD	PAS INC.	P2	same	F2X
432	LA	RENNTECH INC.	R2	same	R2X
433	HD	RENAULT VEHICLES INDUSTRIELS	R3	same	R3X
439	LD	RAYTON-FISSORE NORTH AMERICA	R1	same	R1X
440	LD	ROLLS-ROYCE MOTORCARS LTD.	RR	same	RRX
453	MC	ROSCETTI	RC	same	RCX
454	LD	RUF AUTOMOBILE GMBH	RA	same	RAX
457	LA	ROYALE LIMOUSINE MANUFACTURERS	RL	same	RLX
460	LD	ROVER GROUP LTD.	LR	same	LRX
470	LD	SAAB	SA	same	SAX
	HD	SAAB SCANIA	SS	SA	SAX
471	LD	SAAC CAR COMPANY INC.	S6	same	S6X
472	LA	SALEEN AUTOSPORT	S3	same	S3X
473	LD	SALEEN PERFORMANCE PARTS, INC.	S8	same	S8X
475	LA	SEGUELS SERVICE INC	S2	same	S2X
481	LD	SHELBY AUTOMOBILES INC	SY	same	SYX
487	LD	SLP ENGINEERING	S5	same	S5X
490	LD	MITSUBISHI	MT	same	MTX
	HD	MITSUBISHI	MM	MT	MMX
491	LA	MITSUBISHI MOTOR SALES AMERICA	M3	same	M3X
492	LD	MITSUBISHI MOTORS AUSTRALIA LTD	ML	same	MLX
515	LD	SUPERIOR OF OHIO INC	V1	same	V1X
520	LD	EXCALIBUR AUTOMOBILE	EX	same	EXX
526	LD	TDM TECHNOLOGIES, INC.	T4	same	T4X
527	LA	THOMAS PUGH AND LINDA MCKNIGHT	T3	same	T3X
529	HD	TRANSI-CORP	T5	same	T5X
530	MC	TRIUMPH DESIGNS LTD	TD	same	TDX
534	LA	SPORTS CAR AMERICA PUMA DIVISION	Z4	same	Z4X
540	LA	SUZUKI MOTOR CORPORATION	SK	same	SKX
560	LD	MAZDA MOTOR CORP.	TK	same	TKX
570	LD	TOYOTA	TY	same	TYX
576	LD	NEW UNITED MOTOR MFG INC	NT	same	NTX
579	LD	UTILIMASTER CORP. OF AMERICA	Z1	same	Z1X
581	MC	URALMOTO JSC	YP	same	YPX
582	LD	UNITED STATES COACHWORKS	Y6	same	Y6X

MFR Code	Product	Manufacturer or laboratory	Manufacturer/lab Subcodes		
			<1994	<1997	≥1998
583	LD	US TRADE CORP.	Z2	same	Z2X
590	LD	VOLKSWAGEN	VW	same	VWX
600	LD	VOLVO	VV	same	VVX
603	LD	WALLACE ENVIR.TESTING LAGS. INC	WA	same	WAX
605	HD	VOLVO WHITE TRUCK DIVISION	VT	same	VTX
608	LD	WISCONSIN LIFT TRUCK CORP.	WL	same	WLX
611	MC	WESTWARD INDUSTRIES	WW	same	WWX
614	LD	YUGO AMERICA, INC.	YA	same	YAX
615	MC	YAMAHA	YA	YM	YMX
640	LD	AUDI	AD	same	ADX
645	LD	AMPHI-RANGER OF AMERICA	Y1	same	Y1X
660	LD	FUJI HEAVY IND	FJ	same	FJX
661	SN	FUJI ROBIN INDUSTRIES LTD.	FN	same	FNX
691	LD	LAMBORGHINI	NL	same	NLX
720	HD	WINNEBAGO INDUSTRIALS	WB	same	WBX
728	HD	ASQUITH MOTOR CARRIAGE CO. LTD	A7	same	A7X
730	HD	CATERPILLER	CT	CP	CPX
735	HD	CLARION MOTORS	CA	same	CAX
	MC	CLARION MOTORS	CA	same	CAX
740	HD	CUMMINS	CE	same	CEX
743	HD	DEERE & COMPANY	JD	same	JDX
745	HD	KLOCKNER-HUMBOLT-DEUTZ AG	DZ	same	DZX
747	HD	FLEETWOOD ENTERPRISES	FW	same	FWX
748	HD	GILLIG	GL	same	GLX
750	HD	HERCULES ENGINES	HE	same	HEX
755	HD	IVECO B.V.	VE	same	VEX
760	HD	MACK TRUCKS	MT	MK	MKX
762	HD	MAN NUTZPAHRZEUGE	MN	same	MNX
765	HD	NAVISTAR INTERNATIONAL TRANS.	NV	same	NVX
767	HD	OSHKOSH TRUCK	FT	S7	S7X
770	HD	PERKINS ENGINE COMPANY	PE	PK	PKX
775	HD	ROADMASTER	RM	same	RmX
777	LD,UE,HD,SN,MC,LN,IL,GL,ME				
		JURASSIC PASSENGER CARS27	JP	same	JPX
793	HD	TRANSPORTATION MANUFACTURING COR	T6	same	T6X
795	HD	VIRONEX	VX	same	VXX
802	UE	ANDREAS STIHL	A8	same	A8X
805	UE	BRIGGS & STRATTON	BS	same	BSX
815	LN	DAE HUNG	DE	same	DEX
825	UE	KIORITZ	EH	same	EHX
828	UE	GENERAC CORP	GN	same	GNX
835	UE	HOMELITE TEXTRON	H2	same	H2X

MFR Code	Product	Manufacturer or laboratory	Manufacturer/lab Subcodes		
			<1994	<1997	≥1998
838	UE	HUSQVARNA AB	HV	same	HVX
840	UE	INERTIA DYNAMICS CORP.	N4	same	N4X
845	UE	KOHLER COMPANY	KH	same	KHX
847	UE	KOMATSU ZENOAH AMERICA	KZ	same	KZX
848	LN	KOMATSU LTD.	KL	same	KLX
849	UE	KUBOTA	KB	same	KBX
850	UE	LAWN-BOY	L4	same	L4X
852	UE	LISTER PETTER, INC.	L5	same	L5X
854	SN	MARUYAMA U.S. INC	M4	same	M4X
855	UE	MCCULLOCH CORP.	MH	same	MEX
860	UE	NELSON	NE	same	NEX
865	UE	ONAN CORP	N5	same	N5X
867	SN	SOLO INC	S9	same	S9X
868	UE	POULAN/WEED EATER	PW	same	PWX
869	UE	SHINDAIWA INC	SW	same	SWX
870	UE	TECUMSEH PRODUCTS	TP	same	TPX
871	SN	TANAKA KOGYO CO LTD	T7	same	T7X
872	UE	TELEDYNE TOTAL POWER	T2	same	T2X
885	UE	YANMAR DIESEL ENGINE USA	YD	same	YDX
890	UE	WACKER CORP.	W1	same	W1X
893	SN	WIS-CON TOTAL POWER CORP	WP	same	WPX
901	IL	AUTOMOTIVE TESTING LABS, INC.	01	same	01X
902	IL	ECS LABORATORIES INC.	02	same	02X
903	IL	ENVIRONMENTAL TESTING CORP.	03	same	03X
904	IL	LUCAS ENGINE MANAGEMENT SYSTEMS	04	same	04X
905	IL	ENVIRONMENTAL RESEARCH & DEV. CO	05	same	05X
906	IL	NORTHERN CAL.EMISSIONS LAB.	06	same	06X
907	IL	TESTING SERVICES INC.	07	same	07X
908	IL	COMPLIANCE & RESEARCH SERVICES	08	same	08X
909	IL	AUTOMATED CUSTOM SYSTEMS, INC.	09	same	09X
910	IL	CALIFORNIA ENVIRONMANTAL ENG.	10	same	10X
911	IL	EAGLE PITCHER AUTOMOTIE GROUP	11	same	11X
912	IL	TICKFORD LIMITED	K3	same	K3X
920	GL	COUNTRY OF SWEDEN	SG	same	SGX
980	GL	CALIFORNIA AIR RESOURCES BOARD	80	same	80X
991	GL	EPA CD	91	same	91X
992	GL	EPA EOD	92	same	92X
993	GL	EPA MOD	93	same	93X
994	GL	EPA FOSD	94	same	94X
995	GL	EPA ECTD (obsolete)	95	same	95X
996	GL	EPA RDSD	96	same	96X
997	GL	EPA EPSD	97	same	97X

Appendix H
General Guidelines to Serial Number Locations

TYPICAL SERIAL NUMBER LOCATIONS ON OFF-ROAD EQUIPMENT/ENGINES

Engines

1. Above air filter
2. Above the pulley, or on belt guard if present
3. Fuel pump or manifold
4. Engine block on the side of engine
5. On intake manifold
6. On/near starter
7. On cylinder head
8. On flange (not shown)
9. On valve cover
10. Rear of engine block
11. On compressed air tank (not shown)

Diesel Engine Model – Left Side

Source: Photo taken by Luc Viatour, 2006, www.lucnix.be, http://en.wikipedia.org/wiki/File:Model_Engine_Luc_Viatour.jpg

Diesel Engine Model – Right Side

Source: Photo taken by Luc Viatour, 2006, www.lucnix.be, http://en.wikipedia.org/wiki/File:Model_Engine_B_Luc_Viatour.jpg

Air Compressors

1. On body of compressor motor or engine
2. On cylinder block
3. On draw bar (not shown)
4. On engine bell housing
5. On fan shroud
6. On frame
7. On/near instrument panel (not shown)
8. On/near radiator (not shown)
9. On tank

Boring and Drilling Rigs

1. On boom mount
2. On frame
3. On track
4. On/near instrument panel
5. On/under operator's cab (not shown in this photo)

Source: http://en.wikipedia.org/wiki/File:RC_Drill_Rig_Western_Australia.jpg

Cranes

1. On base plate
2. On bumper
3. On frame
4. On instrument panel (inside cab)
5. On turntable/turntable frame
6. On/below operator's seat (not shown in photo)
7. Operator's cab door (not shown in photo)
8. Outside panel of operator's cab

Crawler Tractors/Dozers

1. Near fuel tank fill (not shown)
2. Near oil fill (not shown)
3. On battery box
4. On firewall (inside cab) (not shown)
5. On frame
6. On tool compartment door
7. On underside of hood (topside is shown)
8. On/inside operator's cab or near cab door
9. On/near instrument panel (inside cab) (not shown)
10. On/near operator's steps

Source: http://en.wikipedia.org/wiki/File:Liebherr_722_Planierraupe_1.JPG

Excavators

1. Between boom cylinders
2. In front or on side of operator's seat
3. Inside the operator's cab on a panel (not shown)
4. Near the boom foot or on the base of the boom (not shown)
5. On hood
6. On main frame
7. On or below the steps
8. On the side of the operator's cab below the window or near the fan
9. On front of cab or outside of the cab on the boom side (To find this, stand between the tracks, facing the front of the excavator and the plate is usually visible somewhere.)
10. Near cab door

Source: http://en.wikipedia.org/wiki/File:Excavator_in_Brittany_France.JPG

Generators

1. On housing
2. Near instrument panel (not shown)

Loaders

1. Near articulation joint
2. On firewall inside operator's cab (not shown)
3. On frame
4. On underside/outside of hood
5. On operator's cab (on the outside, near the door)
6. On/near instrument panel (inside cab) (not shown)
7. On/near lift arm
8. On/near steps
9. On/under battery access door (not shown)

Motor Graders

1. On frame/gooseneck
2. On instrument panel (inside cab) (not shown)
3. Near cab door
4. On engine compartment door

Off-highway Trucks

1. Inside operator's cab (not shown)
2. On bumper
3. On door (not shown)
4. On frame
5. On instrument panel (inside cab) (not shown)
6. Near cab door (not shown)
7. Base of access ladder/steps

Source: Photo taken by Bidgee, http://en.wikipedia.org/wiki/File:Caterpillar_D350D.jpg

Pavers

1. Near operator's platform
2. Near controls or instrument panel
3. On the bin
4. Under operator's seat
5. On side panels near operator's seat
6. Engine compartment cover

Rollers/Compactors
1. Below operator's seat (inside operator's cab) (not shown)
2. Cab/canopy structure
3. On chassis frame (not shown)
4. On side of drum
5. On steps (not shown)
6. On/below operator platform
7. Side panels
8. Steering console/control panel (inside operator's cab) (not shown)
9. Panel near cab door (not shown)

Source: Photo taken by Jan Mehlich, 29 September 2006, http://en.wikipedia.org/wiki/File:Dynapac_CC232.JPG

Rough Terrain Forklifts

1. On control panel (inside operator's cab) (not shown)
2. On fender (not shown)
3. On frame
4. On outside of engine door
5. Outside of operator's cab

Source: Photo taken by Calibas, 14 November 2007, http://en.wikipedia.org/wiki/File:Telescopic_handler2.jpg

Scrapers

1. Inside cab, near controls
2. On bumper
3. On frame/gooseneck
4. On scraper bin/bowl
5. Outside of cab, near door

Source: Photo taken by Bill Jacobus, http://en.wikipedia.org/wiki/File:Scraper.jpg

Skid Steer Loaders

1. On rear panel
2. On frame
3. On/below arm lift
4. On/inside operator's cab
5. On engine cover
6. Near cab entry, viewed from front

Surfacing Equipment

1. Inside engine compartment
2. On conveyor frame
3. On fender
4. On frame
5. On gear box (not shown)
6. On instrument panel (not shown)
7. On tool box (not shown)
8. On/below battery box (not shown)

Tractor/Loaders/Backhoes

1. On frame (not shown)
2. On instrument panel (inside operator's cab) (not shown)
3. On underside/outside of hood
4. On operator's cab
5. On/near arm at pivot point
6. On/near steps
7. Near cab door
8. On battery box

Source: Photo taken by Radomil, http://en.wikipedia.org/wiki/File:Koparko_ladowarka.JPG

-113-

Trenchers

1. On frame
2. On instrument panel (inside operator's cab)
3. Outside of cab
4. Engine compartment, near cab

Source: Photo taken by Ky MacPherson, 25 June 2006, http://en.wikipedia.org/wiki/File:Trencher_2006-06-25.km.jpg

www.ingramcontent.com/pod-product-compliance
Lightning Source LLC
Chambersburg PA
CBHW080642180526
45168CB00008B/3268